André Gerber

Global Analysis of Protein-RNA Interactions

André Gerber

Global Analysis of Protein-RNA Interactions

Discovery of RNA-Protein Networks and Their Implications in Human Disease

Südwestdeutscher Verlag für Hochschulschriften

Impressum/Imprint (nur für Deutschland/only for Germany)
Bibliografische Information der Deutschen Nationalbibliothek: Die Deutsche Nationalbibliothek verzeichnet diese Publikation in der Deutschen Nationalbibliografie; detaillierte bibliografische Daten sind im Internet über http://dnb.d-nb.de abrufbar.

Alle in diesem Buch genannten Marken und Produktnamen unterliegen warenzeichen-, marken- oder patentrechtlichem Schutz bzw. sind Warenzeichen oder eingetragene Warenzeichen der jeweiligen Inhaber. Die Wiedergabe von Marken, Produktnamen, Gebrauchsnamen, Handelsnamen, Warenbezeichnungen u.s.w. in diesem Werk berechtigt auch ohne besondere Kennzeichnung nicht zu der Annahme, dass solche Namen im Sinne der Warenzeichen- und Markenschutzgesetzgebung als frei zu betrachten wären und daher von jedermann benutzt werden dürften.

Coverbild: www.ingimage.com

Verlag: Südwestdeutscher Verlag für Hochschulschriften GmbH & Co. KG
Heinrich-Böcking-Str. 6-8, 66121 Saarbrücken, Deutschland
Telefon +49 681 37 20 271-1, Telefax +49 681 37 20 271-0
Email: info@svh-verlag.de

Approved by: Zürich, ETH, Habil., 2010

Herstellung in Deutschland:
Schaltungsdienst Lange o.H.G., Berlin
Books on Demand GmbH, Norderstedt
Reha GmbH, Saarbrücken
Amazon Distribution GmbH, Leipzig
ISBN: 978-3-8381-1338-8

Imprint (only for USA, GB)
Bibliographic information published by the Deutsche Nationalbibliothek: The Deutsche Nationalbibliothek lists this publication in the Deutsche Nationalbibliografie; detailed bibliographic data are available in the Internet at http://dnb.d-nb.de.

Any brand names and product names mentioned in this book are subject to trademark, brand or patent protection and are trademarks or registered trademarks of their respective holders. The use of brand names, product names, common names, trade names, product descriptions etc. even without a particular marking in this works is in no way to be construed to mean that such names may be regarded as unrestricted in respect of trademark and brand protection legislation and could thus be used by anyone.

Cover image: www.ingimage.com

Publisher: Südwestdeutscher Verlag für Hochschulschriften GmbH & Co. KG
Heinrich-Böcking-Str. 6-8, 66121 Saarbrücken, Germany
Phone +49 681 37 20 271-1, Fax +49 681 37 20 271-0
Email: info@svh-verlag.de

Printed in the U.S.A.
Printed in the U.K. by (see last page)
ISBN: 978-3-8381-1338-8

Copyright © 2011 by the author and Südwestdeutscher Verlag für Hochschulschriften GmbH & Co. KG and licensors
All rights reserved. Saarbrücken 2011

Table of Contents

1. Summary... 3

2. Introduction.. 4
    2.1   Posttranscriptional gene regulation.............................. 4
    2.2.   RNA-binding proteins... 5
    2.3   Non-coding RNA.. 7

3. Global analysis of posttranscriptional gene regulation.............. 9
    3.1   RNA localization and decay....................................... 9
    3.2.   Translational regulation.. 10

4. Systematic mapping of RNA-protein interactions..................... 13
    4.1   RNA-immunopurification Chip analysis (RIP-Chip)............... 13
    4.2   Defining RNA targets for PUF family RNA-binding proteins..... 15
    4.3   Exploring the posttranscriptional regulatory landscape in yeast........... 20
    4.4   Future approaches to study posttranscriptional gene regulation............ 23

5. Emerging principles of the RNA-protein interaction network........ 24

6. RNP networks in human disease.. 28
    6.1   RBP loss-of-function.. 28
    6.2   Mutations in *cis*-acting RNA elements......................... 30
    6.3   RNA gain-of-function and myotonic dystrophy.................. 31
    6.4   Perspectives for drug development and therapy................ 34

7. Concluding remarks... 36

8. References.. 37

Acknowledgements.. 54
Appendix I-VII.. 55

1. Summary

Gene expression is regulated at multiple levels to ensure coordinated synthesis of the cells' macromolecular components. Besides transcriptional regulation, the control of the later post-transcriptional steps has substantial impact on gene expression with widespread implications in physiologically important processes such as development, metabolism, neuronal function, and for cancer progression. On the one hand, posttranscriptional regulation is mediated by RNA-binding proteins (RBPs), which control almost every aspect of RNA's life in a dynamic manner from RNA maturation, quality control, localization, translation, and degradation. On the other hand, mRNAs are post-transcriptionally regulated via physical interactions with small non-coding RNAs. The best characterized class of such RNAs are microRNAs (miRNAs), ~22 nucleotide long RNA molecules that negatively regulate gene expression. Hundreds of RNA-binding proteins (RBPs) and miRNAs are present in eukaryotic genomes, rivaling in number other classes of regulatory molecules such as transcription factors and kinases and thus, suggests and elaborate system for post-transcriptional control that may affect virtually every message in a cell.

Whereas many classical studies explored the molecular function and physiological impact of post-transcriptional regulation on specific mRNA substrates, the recent development of genome-wide analysis tools enables now to study the extend and logic of post-transcriptional gene regulation (PTGR) on a global scale. I have pioneered such 'ribonomic' studies and established methods to affinity isolate RNA-binding proteins and systematically analyzed bound RNAs with DNA microarrays for more than 50 RBPs from yeast, flies, and humans. These studies revealed that RBPs preferentially associate with messages that share common functional and structural attributes suggesting the presence of a highly complex and interweaved post-transcriptional regulatory system. In addition, unraveling the RNA targets for particular RBPs has lead to new insights into their molecular and physiological function. In this habilitation thesis, I summarize some of these investigations and provide an outlook for future research and potential applications for pharmaceutical sciences.

2. Introduction

2.1 Posttranscriptional gene regulation

Regulation of gene expression is fundamental for the coordinated synthesis, assembly and localization of macromolecular structures of cells (Maniatis and Reed, 2002; Orphanides and Reinberg, 2002). Work during the last decades has revealed that this is achieved by a multistep program, which is highly interconnected and tightly regulated at the diverse steps (Orphanides *et al*, 2002). Thereby, research has mainly focused on the first steps of gene expression, namely the transcriptional control mediated by transcription factors (TF) that activate genes by binding to specific DNA promotor sequences and recruit RNA-polymerases for RNA synthesis. Whereas the roles of TFs, chromatin structure and modifications are undisputed, it is now also becoming increasingly recognized that control of the later post-transcriptional steps has substantial regulatory impact with pivotal roles for development, metabolism, neuronal function and aging (Costa-Mattioli *et al*, 2009; Kuersten and Goodwin, 2003; Tavernarakis, 2008).

Post-transcriptional gene regulation has been divided into several subsequent steps (Fig. 1). Upon synthesis of RNA precursors by one of the three RNA polymerases present in eukaryotic cells, these 'early' RNA molecules are immediately bound by a host of RNA-binding proteins (RBPs), forming so-called ribonucleoprotein complexes (RNPs) (Dreyfuss *et al*, 2002; Moore, 2005). The maturation of mRNAs involves several processing reactions occurring in the nucleus: It involves capping, where a tri-methyl guanosine modified cap is added at the 5'-end of messages; the splicing-out of introns, which is performed by a large RNP termed 'spliceosome'; editing, a process where certain RNA bases are modified or inserted/deleted in the transcript changing the coding potential of the messages (Gerber and Keller, 2001); and 3'-end cleavage followed by the addition of long polyadenylic acid (poly[A]) tails, which is important to maintain the mRNA stable and to enhance translation (Dreyfuss *et al*, 2002; Maniatis *et al*, 2002; Moore, 2005) (Fig. 1). After these processing steps, the mRNAs are exported through nuclear pores to the cytoplasm by a variety of export factors (Vinciguerra and Stutz, 2004). Messenger RNAs may further undergo localization to specific subcellular regions by complexes consisting of motor proteins and RBPs or by the signal recognition particle (St Johnston, 2005). Thereby, transport of mRNAs has to be accompanied by translational repression, which is mediated by certain mRBPs (Gebauer and Hentze, 2004; Huang and Richter, 2004). Most mRNAs present in cells (~80% in yeast) are rapidly captured by translation factors and assemble with ribosomes for protein synthesis. Thereby, the translation of messages can be controlled by global means, which involves the post-translational modification of particular translation initiation factors, or through competition. In addition, translation control of specific messages relies on message-specific RBPs. Ultimately, mRNAs undergo exonuclease-mediated degradation by normal, nonsense-mediated (NMD), and nonstop decay (NSD) pathways (Parker and Song, 2004).

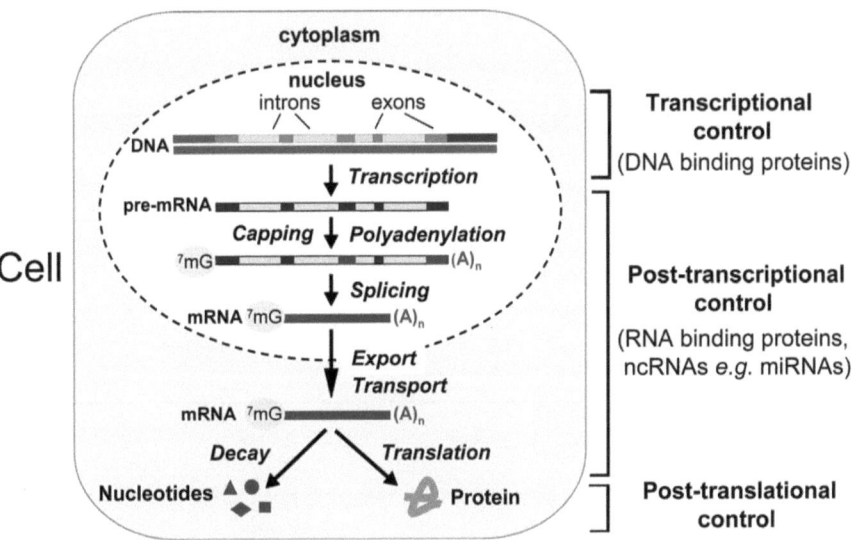

Fig. 1 The multiple-steps of the gene expression program.
RNA transcripts are shown in dark grey; introns in pre-mRNAs are depicted in light grey (adapted with permission from Halbeisen *et al*, 2008).

2.2 RNA-binding proteins

RBPs often contain characteristic RNA-binding domains that specifically interact with sequences or structural elements in the RNA (Table 1). Some well-characterized RNA-binding domains include the RNA-binding domain (RBD, also known as RNP domain and RNA recognition motif, RRM), the K-homology (KH) domain (type I and type II), the RGG (Arg-Gly-Gly) box, the zinc finger motif (ZnF, mostly C-x8-X-x5-X-x3-H), the double stranded RNA-binding domain (dsRBD); the Pumilio/ FBF (PUF or Pum-HD) domain; and the Piwi/Argonaute/ Zwille (PAZ) domain (Auweter *et al*, 2006; Glisovic *et al*, 2008; Lunde *et al*, 2007). Notably, RBPs often contain an array of such RNA-binding motifs, which further increases the specificity and affinity towards the RNA.

Table 1 Examples of eukaryotic RNA-binding domains.

Motif	Characteristics	Proteins	Function/ disorder
RNP motif	RNP consensus sequence composed of two short sequences (8-10aa) separated by ~30 aa babbab fold, four-$\beta\alpha\beta\beta\alpha\beta$ fold, four-stranded antiparallel β sheet packed against two perpendicular oriented α helices.	hnRNP A1, PABP, snRNP U1A, U2AF65	mRNA/ rRNA biogenesis, RNA stability, translation
KH motif	HnRNP K homology domain Core sequence: VIGXXGXXI	FMR-1 Nova Ribosomal proteins	Fragile X-mental retardation Alternative splicing/ brain Translation
dsRBD	70 amino acid region with conserved positions including basic (R, K) and hydrophobic amino acids.	Staufen ADARs RNAse III	Anterior/ posterior axis form. RNA editing Exonuclease
Arginine-rich	10-20 amino acids ariginine-rich sequences	HIV Rev HIV Tat Ribosomal proteins	HIV pre-mRNA export Transcription trans-acitvator Translation
RGG box	20-25 amino acid long, closely spaced RGG repeats interspersed with other, oftern aromatic amino acids.	HnRNP U Nucleolin	mRNA biogenesis rRNA biogenesis
Pumilio	8 repeats each consisting of alpha helices	Pumilio	Anterior/ posterior axis form Neuronal activity
Zinc finger/ knuckle	Appropiately spaced cysteine-histidine residues.	TFIIIA NAB2, TIS11	Pol III transcription/ 5S rRNA mRNA biogenesis/ export mRNA stability

Using these motifs, bioinformatic analyses suggested that eukaryotic genomes encode a large number of RBPs (Anantharaman *et al*, 2002). In yeast, almost 600 proteins (~10% of protein coding genes) are predicted to function as RBPs. In the worm *Caenorhabditis elegans* and in the fly *Drosophila melanogaster*, approximately 2% of the protein coding genes are annotated as RBPs; in humans, more than 1,000 RBPs are annotated representing about 5% of the protein coding genes. However, it is likely that the number of RBPs is even higher, since there are probably other RNA-binding domains that remain to be uncovered. By why do in particular eukaryotes need so many – hundreds and perhaps thousands of RBPs? One possible explanation is that as eukaryotes evolved highly specific post-transcriptional processes to fine-tune gene expression, a concomitant expansion of the number of RBPs occurred. For instance, several RNA-binding domains such as the RRMs underwent drastic amplification during animal evolution, concurrent with the origin of alternative splicing, which allows increasing the genetic diversity with a limited repertoire of genes.

2.3 Non-coding RNAs

Messenger RNA levels are also regulated *via* direct physical interactions with non-coding RNAs (ncRNA). The best-characterized class of such RNAs are small ncRNAs such as the microRNAs (miRNA), ~22-nucleotide (nt) long RNA molecules that negatively regulate gene expression in metazoan (reviewed in Bartel, 2009; Chekulaeva and Filipowicz, 2009; Filipowicz *et al*, 2008). The production of miRNAs is complex and assisted by RBPs that further undergo regulation. It involves the synthesis of primary miRNAs (pri-miRNA) by RNA polymerase II. Pri-miRNAs can encode several miRNAs and are occasionally very long – up to one thousand nucleotides. They from hairpin structures recognized by a nuclear microprocessor complex, consisting of an RNase III enzyme called Drosha, and the dsRBP Pasha/DGCR-8. The complex cleaves pri-miRNAs into ~70 nt long hairpins referred to as precursor miRNAs (pre-miRNAs). Pre-miRNAs are then exported to the cytoplasm by exportin 5 and converted into mature miRNAs by the RNAse III enzyme Dicer. One strand of the duplex is incorporated into the micro-ribonucleoprotein (miRNP) complex – whose main component is a member of the Argonaute (Ago) family. This complex assembles with sequences located mostly in the 3'-UTRs of target mRNAs, and induces changes in subcellular localization, translation efficiency and stability. The rules of miRNA-target recognition are not well understood. One important determinant however is the perfect complementarities between the target site and 7-8 nucleotides at the 5' end of the miRNA (region known as miRNA "seed" (Bartel, 2009). Structural studies with archea Ago proteins indicate that this miRNA region is presented by the RNA-induced silencing complex (RISC) to the target mRNA (Wang *et al*, 2009a). Whether and to what extent nucleotides at the 3' end of the miRNAs are used for interaction with the mRNA target is unknown. Computational analyses have revealed that miRNA target sites often reside in A/U-rich sequence environments towards the boundaries of 3'-UTRs (Grimson *et al*, 2007). To date, almost 1000 different miRNAs have been mapped in vertebrate and based on bioinformatics analyses and experimental evidence, it is estimated that miRNAs regulate more than one third of all human genes – with each miRNA being able to bind up to hundreds of target mRNAs. Consequently, miRNAs have immense potential for PTGR and are thought to be implicated in virtually all biological processes of multicellular organisms, as well as a wide variety of pathological conditions, particularly cancer (Erson and Petty, 2008; Ventura and Jacks, 2009).

Although the current literature is dominated by miRNAs, other classes of ncRNAs exist with strong regulatory potential for gene expression. Strikingly, the recent establishment of high-throughput sequencing methods and computational genomics revealed that up to 90% of the genome is transcribed, producing a bewildering number of different types and forms of RNAs molecules. Among them are long transcripts that, rather than encoding protein, act functionally as RNAs. These long ncRNAs (lncRNAs) - arbitrarily considered to be longer than ~200 nucleotides - were initially found to silence genes on chromatin but nevertheless may also act on mRNA molecules in the cytoplasma (reviewed in Mercer *et al*, 2009). One major emergent theme is the involvement of lncRNAs in regulating the expression of neighboring protein-coding genes. As such, lncRNAs can mediate epigenetic changes by recruiting chromatin remodeling complexes to specific genomic loci. For example, hundreds of lncRNAs are sequentially expressed along the temporal and spatial developmental axes of the human homeobox (Hox) gene loci, where they define chromatin domains with different histone methylation patterns and RNA polymerase accessibility (Rinn *et al*, 2007). The pervasive transcription of enhancers and promoters further anticipates a core role for lncRNA for regulation of transcription. Importantly, proximal promoters can be transcribed into lncRNAs that further recruit and therefore integrate the functions of RBPs into the transcriptional programme, which is exemplified by the repression of cyclin D1 transcription in human cell lines (Wang *et al*, 2008). In this case, DNA damage signals induce the expression of lncRNAs associated with the cyclin D1 gene promoter, where they act cooperatively to modulate the activities of the RNA binding protein TLS. TLS subsequently inhibits the histone acetyltransferase activity of CREB and p300 to silence cyclin D1 expression. The ability of lncRNAs to recruit RBPs to gene promoters largely expands the regulatory repertoire available to the transcriptional program and of RBPs (Wang *et al*, 2008). Lately, the ability of ncRNAs to recognize complementary sequences also allows highly specific interactions that are amenable to regulating various steps in the post-transcriptional processing of mRNAs, including their splicing, editing, transport, translation and degradation. Most mammalian genes express antisense transcripts, which might constitute a class of ncRNA that is particularly adept at regulating mRNA dynamics. For instance, antisense ncRNAs can mask key *cis*-elements in mRNA that are recognized by particular RBPs, through the formation of RNA duplexes. For instance, the *Zeb2* (also called *Sip1*) antisense RNA, which complements the 5' splice site of an intron located in the 5'-UTR of *Zeb2* mRNA, prevents the splicing of that intron, which contains an internal ribosome entry site required for efficient translation and expression of the ZEB2 homeobox protein (Beltran *et al*, 2008). This sets a precedent for ncRNAs in directing the alternative splicing of mRNA isoforms. In conclusion, the pervasive transcription of eukaryotic genomes adds numerous possibilities for regulatory function of ncRNAs and lately to the RBPs with which they are associated. It will be an important field of future research to explore these functions and to unravel the connections to human disease.

3. Global analysis of posttranscriptional gene regulation

The hundreds of RNA-binding proteins (RBPs) and miRNAs encoded in eukaryotic genomes rival in number other classes of regulatory molecules such as transcription factors and kinases and thus, suggest an elaborate system for posttranscriptional control that may affect virtually every message in a cell (Halbeisen *et al*, 2008; Keene, 2007). About 15,000 mRNA molecules are present in each yeast cell during exponential growth and an at least 10-fold larger number in one typical mammalian cell (Hereford and Rosbash, 1977). This vast numbers raise questions about the extent to which the location, activity, and fates of RNA populations are coordinated and about post-transcriptional mechanisms that might mediate their coordinated regulation (Hieronymus and Silver, 2004; Mata *et al*, 2005).

Whereas many classical studies explored the mechanistic and physiological impact of post-transcriptional regulation by RBPs/miRNAs on specific mRNA substrates, the development of genome-wide analysis tools enables the study of post-transcriptional gene regulation on a global scale (Halbeisen *et al*, 2008). A link to an in depth review of the studies that applied this "systems" approach to unravel the extent and logic of RNA localization, RNA decay and mRNA translation is given in Appendix I (Halbeisen *et al*, 2008).

3.1. RNA localization and decay

Global RNA-localization studies revealed unexpectedly large numbers of mRNA molecules that are transported to specific subcellular compartment for local translation. A classical example involves the localization of 24 messages to the bud-tip of dividing yeast cells including Ash1 mRNA, coding for a transcriptional repressor that prevents mating-type switching in the newly born daughter cell (Shepard *et al*, 2003). Microarray studies with RNA obtained from mammalian dendrites, revealed hundreds of messages that are possibly transported in neurons, allowing local translation in response to action potentials (Bramham and Wells, 2007; Matsumoto *et al*, 2007). Lately, an extensive in situ hybridization study in *Drosophila* developing embryos revealed that more than 70% of the 3,000 or so analyzed mRNA transcripts are located to specific regions (Lecuyer *et al*, 2007). These and other data challenges our "trivial" view of RNAs that are randomly distributed in cells and raise the impression that possibly most mRNAs are directed to specific regions in the cells.

Evidence for coordinated post-transcriptional regulatory programs has also been gained from global studies on mRNA decay. Messenger decay rates were measured for each transcript after RNA polymerase inactivation in bacteria (Bernstein *et al*, 2002b), yeast (Grigull *et al*, 2004; Wang *et al*, 2002), plants (Gutierrez *et al*, 2002), and human cell culture (Yang *et al*, 2003). One intriguing result from these investigations was that mRNA half-lives often correlate among components of macromolecular complexes or among members of the same functional class - defining decay 'regulons'. For example, transcripts encoding metabolic proteins generally have long half-lives, whereas transcripts encoding transcription factors are relatively unstable (Wang *et al*, 2002). Moreover, similar mRNA-turnover patterns can be found among orthologous genes in yeast and human indicating evolutionary conserved programs of RNA stability control. Such decay programs may also apply to particular subcellular compartments as seen for IRE1 from *Drosophila*, a protein activated during the unfolded protein response in the endoplasmatic reticulum, which directs the decay of a specific subset of mRNAs, many of them coding for plasma-membrane proteins (Hollien and Weissman, 2006). In conclusion, these studies underline the importance of RNA decay in the control of mRNA levels and furthermore, strongly suggest the presence of specific RNA turnover programs.

3.2. Translational regulation

Several laboratories have used DNA microarray technology to perform genome-wide analysis of mRNAs in polysomes to investigate global aspects of translational regulation (reviewed in Beilharz and Preiss, 2004; Halbeisen *et al*, 2008; Rajasekhar and Holland, 2004). In yeast, it appears that translational and transcriptional programs are tightly coupled such as regulation at the translational level often reflects a magnification of the transcriptional activity - an effect that has been termed 'potentiation' (Preiss *et al*, 2003). To gain a comprehensive view of the post-transcriptional regulatory system, its dynamics, and the coordination between different layers of the gene expression program, we have performed an own comprehensive analysis comparing alterations of global transcript levels (transcriptome) with corresponding ribosome associations (translatome) upon changing environmental conditions e.g. after treatment of cells with drugs (Appendix II; Halbeisen and Gerber, 2009a). This analysis was further motivated by the fact that changes of mRNA levels do not necessarily relate to the encoded proteins, as protein levels will also rely on differential recruitment of mRNAs to translating ribosomes and on protein decay (Gygi *et al*, 1999; Lu *et al*, 2007).

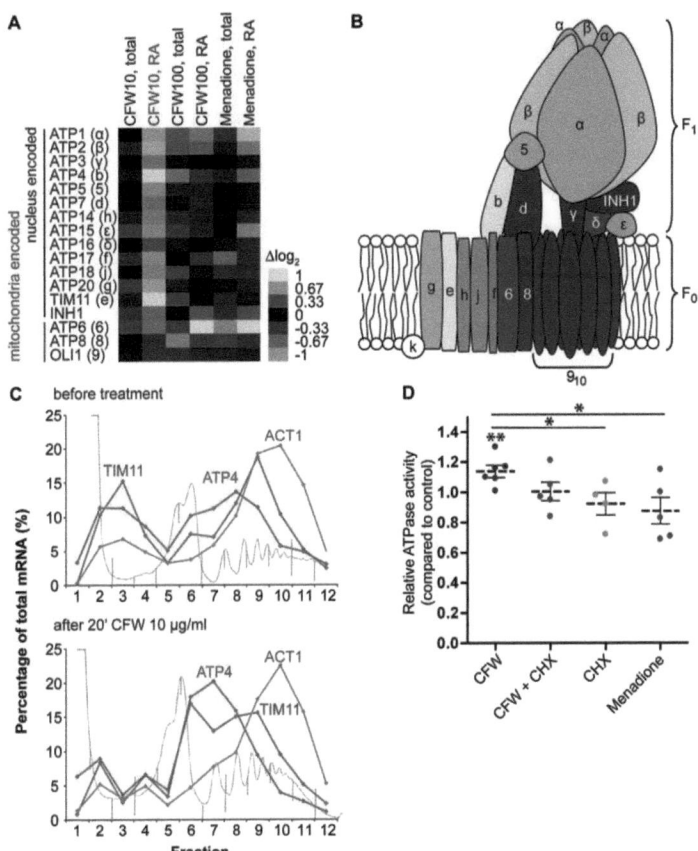

Fig. 2 Enhanced ribosome association and activity of the mitochondrial F1F0-ATPase by low doses of Calcofluor-white (CWF).
(A) Heat map representing relative changes of expression of 17 mitochondrial F1F0-ATPase components after mild stress treatments (white-black scale). Total: changes of steady-state mRNA levels, RA: change of ribosome associations. The gene names are indicated to the left, with the name of the corresponding subunit in parentheses. No changes were seen after treatment of cells with Menadione, a component that induces genotoxic stress.
(B) Structure of the mitochondrial F1F0-ATPase. Each component is colored according to changes in ribosome associations depicted in (A). The membrane-associated part (F0) uses a proton motive force to mechanically drive the soluble part (F1) that exhibits ATPase activity.
(C) Distribution of TIM11, ATP4, and ACT1 mRNAs in polysomal gradients obtained from untreated cells (upper panel), and from cells treated with CFW (lower panel). RNA was isolated from each fraction of the polysomal profile and quantified by RT-qPCR (see Materials and Methods). The mRNA level in each

fraction was calculated as a percentage of the total; data and absorbance (254 nm) profiles from representative experiments are plotted. ACT1 is as a negative control mRNA that was not expected to alter ribosome association.

(D) Relative mitochondrial ATPase activity of drug-treated versus untreated control cells. Cells were treated with CFW (CFW) for 20 min; pre-incubated for 10 min with CHX prior to addition of CFW (CHX+CFW); or treated with CHX or menadione. The activity of purified mitochondria was normalized to the untreated control sample. Average activities are indicated with a dashed line, standard errors of the mean (SEM) with continuous lines. Single dots represent biologically independent experiments (double asterisks [**] indicate $p < 0.01$; a single asterisk [*] indicates $p < 0.05$). (Figure reproduced from Halbeisen et al, 2009a; a link to a colored version of this figure can bee found in Appendix II).

To analyze the relation between transcriptome and translatome, we have first established a novel approach to rapidly access the 'translatome', involving the affinity-purification of endogenously formed ribosomes and the analysis of associated mRNAs with DNA microarrays (Halbeisen et al, 2009b). Using this method, called ribosome-affinity purification (RAP), we have then compared changes in total mRNA levels (transcriptome) with changes in the translatome after the application of different conditions of cellular stress to yeast cells (Appendix II; Halbeisen et al, 2009a). We found that changes of the transcriptome correlate well with those of the translatome after application of "severe" stresses that stop cell growth. However, this correlation was generally lost under mild stresses that do not affect cell growth. In this case, remodeling of gene expression was mainly executed at the translational level – redirecting messages in and out of ribosomes as we have further validated for components of the mitochondrial ATPase (Fig. 2). In this case, eight of the 14 nuclear-encoded messages coding for components of the mitochondrial ATPase showed increased ribosome association but total mRNA levels remained unchanged after application of low doses of Calcofluor-white, a drug that selectively induces cell-wall stress. These data suggested that low doses of CFW specifically induce the "relocalization" of those mRNAs to ribosomes that code for the mitochondrial ATPase complex (Fig. 2). Therefore, our study showed that protein synthesis is under selection to modulate the expression of functionally related messages coding for macromolecular complexes to rapidly adapt to minor changes in cellular needs.

4. Systematic mapping of RNA-protein interactions

4.1. RBP-immunopurification Chip analysis (RIP-Chip)

The application of genome-wide analysis tools such as DNA microarrays allows now the systematic exploration of the post-transcriptional system (Halbeisen *et al*, 2008). One particular application that I have been focusing on involves the systematic identification of RNAs that copurify with endogenously expressed RBPs – applying a method now commonly referred to as RBP immunoprecipitation-microarray (RIP-Chip) profiling (Fig. 3). This method involves the affinity purification of tagged RNP complexes from cellular extracts. Alternatively, antibodies that selectively recognize an epitope of a constituent protein can also been used to immune-purify RNP complexes. The purified RNPs are then dissociated into proteins and RNAs, and the identities of RNAs are determined on a global-scale with DNA microarrays. Likewise, protein components can be identified with mass-spectrometry or specific proteins are identified with immunoblots.

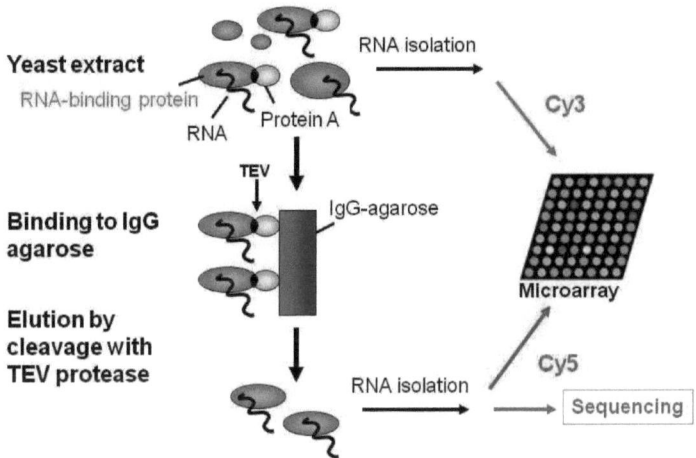

Fig. 3 RIP-Chip to systematically identify RNA targets of RBPs.
RBPs are immunopurified or affinity-purified via a tag from cellular extracts. Total RNA isolated from cell extracts, and the RNAs associated with the particular RBP under study are labeled with different fluorescent dyes and competitively hybridized to DNA microarrays. The fluorescence ratio for each locus reflects its enrichment by affinity for the cognate protein. In the future, the application of novel sequencing technologies will certainly increase the sensitivity and reproducibly of the assay (Figure adapted from Gerber *et al*, 2004).

RIP-Chip was first established in Jack Keene's lab to study RNAs associated with RBPs in human cancer cells using antibodies against endogenous proteins (Tenenbaum *et al*, 2000). Diverse groups have then further developed and modified the RIP-Chip protocol, such as the affinity isolation of recombinant or endogenously expressed epitope-tagged RBPs. RIP-Chip has been employed in diverse species including yeast (Gerber *et al*, 2004; Hieronymus and Silver, 2003; Hogan *et al*, 2008; Inada and Guthrie, 2004), flies (Gerber *et al*, 2006), worms (Roy *et al*, 2002), plants (Schmitz-Linneweber *et al*, 2005), and in various cell lines (Galgano *et al*, 2008; Lopez de Silanes *et al*, 2004; Lopez de Silanes *et al*, 2005; Morris *et al*, 2008; Penalva and Keene, 2004; Townley-Tilson *et al*, 2006) (for a more comprehensive list of references for RIP-Chip experiments performed in various tissues and species, see Morris *et al*, 2010). RIP-Chip has also been used by us and others to identify the messages that are actively transported by RNP-motor complexes to particular subcellular regions (Elson *et al*, 2009; Shepard *et al*, 2003; Takizawa *et al*, 2000). For instance, in the yeast *Saccharomyces cerevisiae,* mRNAs that are asymmetrically distributed between mother and daughter cells were revealed upon affinity-purification of the RNA transport components She2, She3, and the myosin motor protein Myo4, followed by DNA microarray analysis combined with a secondary GFP-based RNA reporter assays to visualize RNAs in living cells (Takizawa *et al*, 2000; Shepard *et al*, 2003). In addition to the known bud-localized Ash1 and Ist2 messages, this analysis revealed 22 additional polarized mRNAs, suggesting the existence of widespread cytoplasmic mRNA localization in yeast (Shepard *et al*, 2003).

Besides its application to specific RNP complexes, RIP approaches have also been applied to more 'general' RNA-binding proteins to identify the messages expressed in particular tissues or cell-types. For instance, tagged poly(A) binding protein (PABP), which binds to the poly(A) tail of cytoplasmic mRNAs, has been expressed within cells or tissues of interest using specific promoters to identify muscle- or ciliated sensory neuron-specific transcripts in the worm *C. elegans* (Kunitomo *et al*, 2005; Roy *et al*, 2002), or in photoreceptor cells of the fly *Drosophila melanogaster* (Yang *et al*, 2005). Likewise, tagged ribosomal proteins have been pulled-down to isolate actively translating ribosomes to define gene expression profiles of specific neuronal cells types in the central nervous system of mice (Heiman *et al*, 2008), and to study reactions of the translatome upon diverse stress conditions in yeast (Halbeisen *et al*, 2009a). Finally, the RIP approach has also been used to identify targets that potentially undergo miRNA dependent regulation by profiling RNA and proteins that are bound by Ago proteins (Beitzinger *et al*, 2007; Hendrickson *et al*, 2008; Hendrickson *et al*, 2009; Karginov *et al*, 2007; Landthaler *et al*, 2008; Zhang *et al*, 2007; Zhang *et al*, 2009). The comparison of Ago-associated mRNAs in wild-type and miRNA mutants or over-expressing cells further provides a tool to decipher miRNA-specific targets.

A major advantage of RIP is that it is a straightforward protocol that allows for the concomitant identification of RNA and protein components of RNPs. Drawbacks of RIP are concerns that during the procedure certain RNAs or proteins may fall-off and others associate with RNP complexes (Mili and Steitz, 2004). However, how often this occurs has not been conclusively answered and may depend on the RBP under investigation. For the isolation of unstable RNP complexes, modifications of the RIP procedure have been proposed to better preserve RNP complex stability and integrity (Baroni *et al*, 2008). RNPs have also been cross-linked either by UV or with chemicals prior to RIP-Chip (San Paolo *et al*, 2009). A more elaborate but very specific UV-light cross-linking-immunoprecipitation (CLIP) protocol has been developed by Ule, Darnell and colleagues, which allows for the fairly precise mapping of RNA-protein or RNA-RNA interactions sites on transcripts (Chi *et al*, 2009; Jensen and Darnell, 2008; Ule *et al*, 2005; Ule *et al*, 2003; Wang *et al*, 2009b; Yeo *et al*, 2009). The drawbacks of this method are, however, a more complicated experimental set-up, and it hardly allows for the analysis of co-purifying proteins. In addition, UV irradiation may chemically and physically alter the RNP structures and cause some sequence bias due to the unequal photo reactivity between bases and amino acids (Gaillard and Aguilera, 2008). Recently, the Tuschl and Zavolan labs have developed an improved CLIP approach allowing transcriptome-wide determination of cellular RBPs and miRNPs binding-sites at high resolution (Hafner *et al*, 2010). In this approach, called PAR-CLIP, the crosslinked sites are revealed by thymidine to cytidine transitions in the cDNAs prepared from immunopurified RNPs of 4-thiouridine-treated cells. To validate their approach, they determined the binding sites and regulatory consequences for several intensely studied RBPs and miRNPs, including PUF and Ago proteins, revealing that these factors bind thousands of sites containing defined sequence motifs (Hafner *et al*, 2010).

In line of these recent developments, the future implementation of high-throughput sequencing (HITS) to identify bound RNAs will certainly increase the reproducibility and the sensitivity of these assays (Shendure and Ji, 2008), possibly enabling to quantify the number of message partitioned into distinct subcellular compartments (Chi *et al*, 2009; Fox *et al*, 2009; Licatalosi *et al*, 2008).

4.2. Defining RNA targets for PUF family RNA-binding proteins

A prime example for the coordination of functionally related transcripts by RBPs is demonstrated by the Pumilio-Fem-3 binding (PUF) proteins, which I have initially focused on and systematically identified targets in yeast, flies, and human cells (Appendix III-V; Gerber *et al*, 2004; Gerber *et al*, 2006; Galgano *et al*, 2008). In the following, I will briefly introduce these proteins and highlight some of the major findings that we gathered from mapping their RNA targets on a global scale.

The PUF proteins comprise an evolutionarily conserved family of RBPs that are implicated in various physiological processes (Bernstein *et al*, 2002a; Spassov and Jurecic, 2003). They are defined by the presence of an RNA-binding domain, termed Pumilio-homology domain (Pum-HD), which consists of eight repeats, each of which makes contact with a different RNA base (Edwards *et al*, 2001; Miller *et al*, 2008; Wang *et al*, 2001). PUF proteins bind to an RNA element that comprises a core 'UGUR' tetranucleotide followed by 3'-UTR sequences that vary among PUF proteins. In concert with other factors, PUFs repress gene expression by inhibiting translation or promoting decay (Goldstrohm *et al*, 2006; Goldstrohm *et al*, 2007; Kadyrova *et al*, 2007; Olivas and Parker, 2000). The study of PUF proteins in diverse model organisms revealed widespread roles for these proteins in embryonic development, stem-cell maintenance and neurogenesis (Spassov *et al*, 2003; Wickens *et al*, 2002). For instance, in the fruit fly *Drosophila melanogaster*, Pumilio (Pum) is required for proper anterior/posterior patterning during early embryogenesis by repression of the translation of *hunchback* mRNA (Murata and Wharton, 1995). Furthermore, Pum is also involved in the development and migration of primordial germ cells (Asaoka-Taguchi *et al*, 1999; Forbes and Lehmann, 1998; Lin and Spradling, 1997), and it may be implicated in long-term memory formation and neuronal excitability (Dubnau *et al*, 2003; Menon *et al*, 2004; Muraro *et al*, 2008). The six yeast *Saccharomyces cerevisiae* PUF proteins (Puf1p – Puf6p) regulate aging, mating-type switching and mitochondrial function (Garcia-Rodriguez *et al*, 2007; Tadauchi *et al*, 2001; Wickens *et al*, 2002). Two paralogous PUF proteins exist in humans, termed Pumilio homolog 1 (PUM1) and Pumilio homolog 2 (PUM2). PUM1 and PUM2 are often co-expressed in diverse tissues suggesting that they may occasionally act redundantly (Macdonald, 1992; Moore *et al*, 2003; Spassov *et al*, 2003). Based on few studies investigating PUM2 function, it was assumed that mammalian PUFs have physiological roles analogous to the non-vertebrate homologs. In germ cells, PUM2 interacts with deleted in azoospermia (DAZ), DAZ-like (DAZL) proteins, and the meiotic regulator BOULE (BOL), which are RBPs that function in early germ line stem cells (Moore *et al*, 2003; Urano *et al*, 2005). Moreover, mouse *Pum2* mutants have smaller testes, although fertility seems not to be affected (Xu *et al*, 2007), suggesting a role for *Pum2* in the maintenance of germline stem cells. PUM2 also negatively regulate the expression of MAPK1 (mitogen-activated protein kinase 1, ERK2) and MAPK14 (mitogen-activated protein kinase 14) in human embryonic stem cells and in the *C. elegans* germline. Recent evidence suggests additional roles of mammalian PUM2 in neurons by regulating dendrite morphogenesis and synaptic function (Vessey *et al*, 2010; Vessey *et al*, 2006; Zhong *et al*, 2006).

I started the systematic identification of PUF targets in yeast and thereby established a robust RIP-Chip procedure (Appendix III; Gerber *et al*, 2004) The analysis revealed that each of the five yeast PUF proteins associated with distinct groups of 40 to 220 different mRNAs with striking common themes in the functions and subcellular localization of the proteins they encode: Puf3p binds nearly exclusively to cytoplasmic mRNAs that encode mitochondrial proteins; Puf1p and Puf2p interact preferentially with mRNAs encoding membrane-associated proteins; Puf4p preferentially binds mRNAs encoding nucleolar ribosomal RNA processing factors; and Puf5p is associated with mRNAs encoding chromatin modifiers and components of the spindle pole body (Fig. 4). The results were further corroborated by the identification of distinct sequence motifs in the 3'-untranslated regions of the mRNAs bound by Puf3, Puf4, and Puf5 proteins. A physiological relation between Puf3p and its mRNA targets has also been observed - as suggested from its association with mRNA-encoding mitochondrial proteins, *puf3* mutant cells showed a slow-growth phenotype on non-fermentable carbon sources indicative of a functional connection to mitochondrial physiology (Gerber *et al*, 2004). In agreement with these findings, more recent studies from us and other laboratories showed that Puf3 could guide certain mRNAs coding for mitochondrial proteins to mitochondria (Eliyahu *et al*, 2010; Saint-Georges *et al*, 2008), and that it plays a role for mitochondrial biogenesis (Garcia-Rodriguez *et al*, 2007).

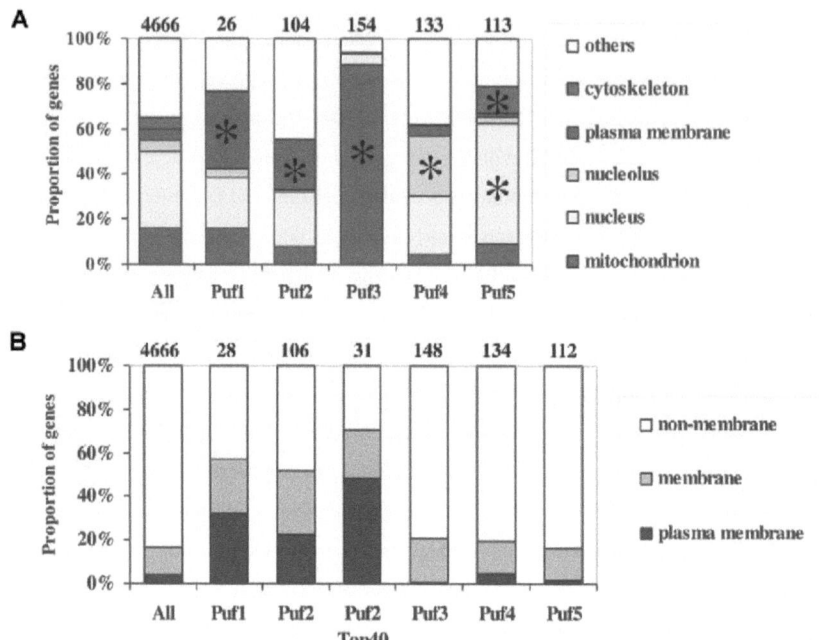

Fig. 4 Classification of mRNAs interacting with yeast Puf Proteins.
(A) Column charts showing compartmentalization of characterized gene products encoded by the Puf targets. The same compartments are shown for the entire genome in the columns designed "All" (YPD, May 2003). The number of genes represented in the charts is indicated on the top of columns. An asterisk indicates classes with p values of less than 0.001. (B) Fraction of membrane-associated gene products among the Puf targets. We classified the targets by combining both GO and YPD annotations (May 2003). "Plasma membrane" (light blue) is a subpopulation of the total membrane-associated proteins (blue). Soluble cytoplasmic or nuclear proteins were classified as "non-membrane." "All" refers to the genome-wide compartmentalization of characterized genes, and respective numbers were retrieved from YPD. "Puf2 Top 40" refers to the 40 highest enriched Puf2p targets and equals the total number of Puf1p targets (Figure adapted from Gerber *et al*, 2004).

Genome-wide identification of RNAs associated with the orthologous PUF protein from *Drosophila melanogaster*, called PUMILIO, revealed distinct clusters of mRNAs in embryos and in ovaries of adult flies (Appendix IV; Gerber *et al*, 2006). More than 1,000 messages were significantly associated with the protein. Subgroups of these Pum-associated mRNAs had commonalities, such as function in the anterior-posterior patterning system, and the subunits of the vacuolar H-ATPase (Fig. 5). A characteristic sequence motif was present in 3'-UTRs of

PUMILIO-bound mRNAs resembling the one previously identified for the yeast Puf3 protein. Hence, the data obtained from the yeast and *Drosophila* studies provided an additional source for considering their evolution. For instance, conservation of amino acid residues in the RNA-binding domain (the PUM-homology domain) between homologous PUF proteins correlated with identified core motifs in 3'-UTR of mRNA targets. However, the proteins encoded by the mRNA targets appeared not to be particularly conserved. This discordance suggested that acquisition or loss of RBP binding motifs in UTRs of genes may provide a surprisingly fluid evolutionary mechanism to modify post-transcriptional regulatory connections (Gerber *et al*, 2006).

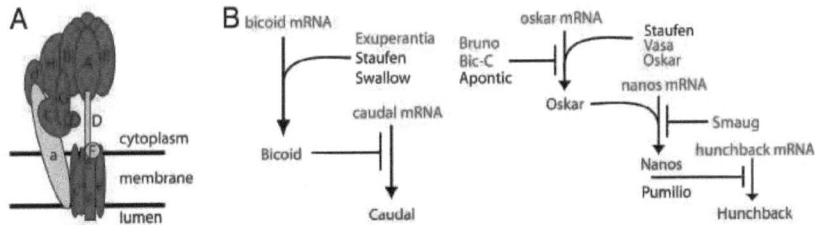

Fig. 5 mRNAs associated with Pum encode proteins of specific macromolecular complexes and regulatory pathways.
(A) Subunits of the vacuolar ATPase. Dark grey, subunits whose mRNAs associated with Pum; gray, subunits whose mRNAs were not enriched with a FDR of <1%. Subunits of the V1 domain are labeled with capital letters: A, *Vha68*; B, *Vha55*; C, *Vha44*; D, *Vha36*; E, *Vha26*; F, *Vha14*; G, *Vha13*; H, *VhaSFD*. Subunits of the V0 domain are indicated by small letters: a, *Vha100*; c, *Vha16*; c, *PPA1*; e, *VhaM9.7*; d, *VhaAC39*. (B) Components of anterior–posterior patterning systems associated with Pum in embryos and/or adults. Messenger RNAs and associated with Pum are shown in grey. Proteins whose mRNAs were not found to be associated are in black (Figure reproduced from Gerber *et al*, 2006).

Lately, we also defined an overlapping set comprised of more than 1,000 messages that were reproducibly associated with human PUM proteins in cancer cells, many of them encoding functionally related proteins that act in medically relevant pathways such as angiogenesis or Ras signaling (Appendix V; Galgano *et al*, 2008). As previously seen with yeast and *Drosophila* targets, a very similar consensus sequence element was highly enriched in the 3'-UTR of targets and confers binding to human PUM proteins (Fig. 6). Strikingly, we found an extensive overlap of PUF mRNA targets with evolutionarily conserved miRNA binding sites offering the possibility for functionally relevant localization of PREs downstream the miRNA sites. Moreover, high-confidence miRNA binding sites were significantly enriched in the 3'-UTRs of experimentally determined PUM1 and PUM2 targets. These results suggested some cross-talk between PUM proteins and miRNA regulatory system, and we are currently analysing this relation with global and specific approaches (Kanitz and Gerber, unpublished results).

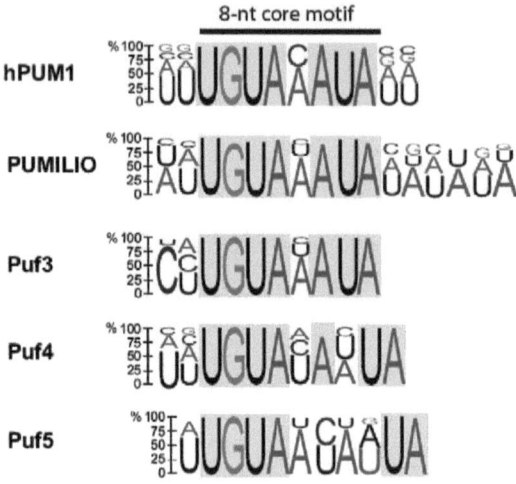

Fig. 6 PUF consensus motif in 3'-UTR sequences associated with PUM1, *Drosophila* Pum and yeast Puf3, Puf4 and Puf5 proteins.
Height of the letters indicates the probability of appearing at the position in the motif. Nucleotides with less than 10% appearances were omitted (Figure adapted from Galgano *et al.* 2008).

4.3 Exploring the post-transcriptional regulatory landscape in yeast

To survey the structure and extend of the proposed coordinative post-transcriptional network, we have by now analyzed the sets of RNAs specifically associated with more than 40 regulatory RBPs in yeast (Appendix VI; Hogan *et al*, 2008). As previously seen for Puf proteins, many RBPs bound mRNAs, which protein products share identifiable functional or cytotopic commonalities (Fig. 7). We observed connections between the RNAs associated with RBPs and the phenotype of RBP-mutant cells, unexpected binding of RBPs to non-coding RNAs, and combinatorial binding of previously unrelated RBPs to the same RNAs. We identified specific sequences or predicted structures significantly enriched in target mRNAs for 16 RBPs. These potential RNA-recognition elements were diverse in sequence, structure, and location: some were found predominantly in 3-untranslated regions, others in 5-untranslated regions, some in coding sequences, and many in two or more of these features. Although this study only examined a small fraction of the universe of yeast RBPs, 70% of the mRNA transcriptome had significant associations with at least one of these RBPs and, on average, each distinct yeast mRNA interacted with three of the RBPs, suggesting the potential for a rich, multidimensional network of regulation that may even outperfom better characterized transcriptional regulatory networks (Hogan et al, 2008).

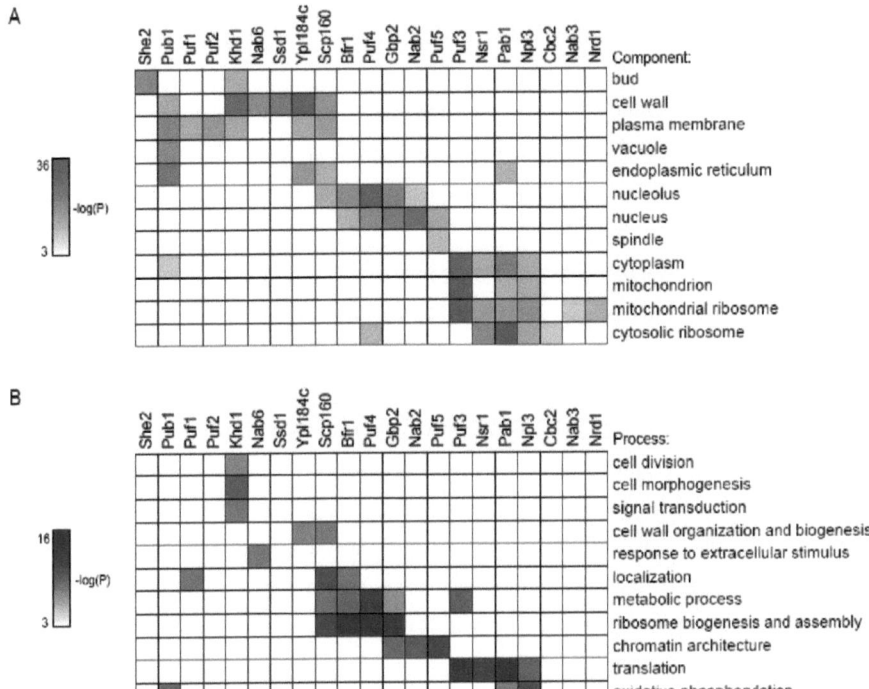

Fig. 7 RBPs bind mRNAs encoding functionally and cytotopically related proteins.

(A) Enrichment of "cellular component" GO terms (rows) in target sets (1% FDR) of RBPs (columns). The significance of enrichment of the GO term is represented as a heat map (scale is to the left of the figure) in which the intensity corresponds to the negative log10 p-value, calculated using the hypergeometric density distribution function and corrected for multiple hypothesis testing using the Bonferroni method. Only a subset of significantly enriched GO terms are shown. RBPs whose targets are significantly enriched ($p < 0.01$) for at least one "cellular component" or "biological process" GO term are shown. (B) Same as in (A), except for "biological process" GO terms (Figure reproduced from Hogan et al, 2008).

To address the combinatorial arrangement of RBPs and the fate of RBP targets, I also studied selected RNA-binding proteins in more detail. In collaboration with Kenji Irie (Tsukuba University, Japan), we studied the targets and functional implication of Khd1p, a yeast protein known to repress translation of *ASH1* mRNA during actin-dependent transport to bud-tips. Applying RIP-Chip, we found that Khd1p binds to hundreds of mRNAs, many of them encoding membrane associated proteins (Hasegawa *et al*, 2008). The combination of bioinformatics, RNA localization, and *in vitro* RNA-binding assays further revealed that Khd1p binds to CNN repeats in coding regions of mRNA targets. Interestingly, we found that Khd1p appears to differentially affect the fates of its mRNA targets; whereas some messages are translationally repressed such as Ash1 mRNAs, other messages become stabilized upon interaction with Khd1p like Mtl1; reflecting the redundant structure of post-transcriptional regulatory systems (Hasegawa *et al*, 2008).

The discovery that most regions of the genome are actively transcribed into non-coding RNAs has dramatically increased the interest into their function and regulation. Recent data has shed light on the molecular machinery that promotes the decay of such transcripts – the so-called TRAMP (Trf4/5-Air1/2-Mtr1) complex degrades aberrant and short-lived RNAs (including "cryptic unstable transcripts" or CUTs) in the budding yeast *S. cerevisiae*. Trf4p and Trf5p are non-canonical poly(A) polymerases that are alternative subunits of these complexes. They add short poly(A) tails to their substrate RNAs that function as landing pads for the exonucleases of the nuclear exosome mediating RNA decay (e.g. of aberrant tRNAs). Although alternate compositions of TRAMP complexes have been characterized by several groups, the RNA substrate specificities and the processes controlled by them have not been analyzed in detail. In collaboration with Prof. Walter Keller's lab (Biozentrum, University of Basel), we have performed a first systematic comparison of the RNA targets and functional implications of different TRAMP complexes (Appendix VII; San Paolo *et al*, 2009). To this end, we combined genome-wide analysis tools (microarray profiling of mutants and establishing a cross-linking immunopurification method X-RIP-Chip), with classical genetic and cell-biological methods to obtain a comprehensive picture of the targets and actions of the TRAMP4 and TRAMP5 complexes in budding yeast. The three key results of this study were: First, we found that Trf4p and Trf5p target distinct groups of RNAs for degradation. Previous studies suggested largely overlapping targets based on genetic studies showing that cells bearing single deletions of either of the two genes are viable, whereas the double deletions are lethal suggesting strong functional redundancy. Second, we found that most RNAs that were altered by the *trf4* deletion were restored to wild type levels by the overexpression of *TRF4(DADA)*, a mutant in which two essential aspartate residues in the polymerase's active site had been converted to alanine, leading to the complete loss of polyadenylation activity. This showed that the polyadenylation activity of Trf4p is dispensable *in vivo* and suggests that most RNA targets of the TRAMP4 complex do not require a poly (A) tail in order to be digested by the nuclear exosome. This finding was unexpected as it was generally thought that the protein needs the polyadenylation activity to initiate RNA decay by the exosome. Third, we identified new roles for the TRAMP complexes in RNA metabolism. For instance, our data provided evidence for the implication of TRAMP complexes in spliced-out intron decay, a process that to our knowledge has not been analyzed in great detail yet. In addition, we showed that disruption of *trf4* causes severe shortening of telomeres, and we provided experimental evidence that *TRF4* functions in telomere maintenance. In conclusion, these results provided a first map for overlapping yet distinct functional specificities of TRAMP complexes, and demonstrate strong connections between RNA surveillance and other RNA-related processes. (San Paolo *et al*, 2009).

4.4 Future approaches to study posttranscriptional regulation

The ability to isolate native ribonucleoprotein particles (RNPs) is fundamental in the study of posttranscriptional gene expression. As outlined above, to date, most laboratories have employed RIP or CLIP procedures to systematically map the RNA targets for RBPs, the binding sites and/or identified RNP components with mass-spectrometry (MS). Therefore, only the development of genomics tools enabled approaching the RNA-protein interaction network on global scale. In the future, one aim will certainly be to reduce the sample size - possibly allowing single-cell analysis - and requires increasing the sensitivity and specificity of these assays. The employment of next generation sequencing tools will certainly be very helpful to get a more robust and quantitative analysis of the RNA targets, including the identification of rare and novel RNA molecules/species (Shendure *et al*, 2008) (see also comments above in section 4.1).

On the other hand, little information is available regarding the full complement of RBPs and miRNAs associated with a specific RNA. This becomes even more important in light of the rapidly increasing number of ncRNAs – in particular the lncRNAs – playing important roles for gene expression control in cell-differentiation and development. Such information will also be important to unravel the combinatorial regulation of mRNAs by multiple RBPs, and how the assorted RBPs change with environmental cues. Several labs have therefore begun to develop methods to identify RBPs (and possibly also miRNAs) that are associated with ncRNAs or specific mRNAs (Butter *et al*, 2009; Hartmuth *et al*, 2004; Hogg and Collins, 2007; Srisawat and Engelke, 2001; Vasudevan and Steitz, 2007; Windbichler and Schroeder, 2006). For this purpose, the 3'UTR (or other regions) of transcripts are intrinsically tagged with RNA tags, called aptamers, which allow selection of tagged transcript by affinity purifications. For example, the streptavidin (S1) tag has been successfully used to purify proteins bound to AU-rich sequences (ARE elements) from human cells (Butter *et al*, 2009; Vasudevan *et al*, 2007). In our experience, the reported procedures are not very robust yet - possibly due to the low copy number of most messages. In this respect, it would be interesting to further develop RNA tags allowing the rapid and robust purification of all kinds of transcript from cells.

Besides the analysis of the targets for RBPs/miRNAs, the combination of genomic and quantitative proteomics should permit the quantitative description of mRNA translation and degradation. Novel large-scale proteomics approaches enable now the quantitative measurement of hundreds of proteins in parallel (Ong and Mann, 2005; Schiess *et al*, 2009). Concomitant analysis of the changes in the RNA and protein levels upon RBP depletion or overexpression could therefore provide global information of the different status of mRNAs and of their downstream effects (Baek *et al*, 2008; Selbach *et al*, 2008). We have conducted such an analysis in collaboration with Ruedi Aebersold's group at the ETH Zurich, where we analyzed concomitant changes of mRNAs and proteins upon overexpression of Gis2, a zinc-finger containing RBP from yeast (see chapter 5; Scherrer *et al*, 2011).

In the end, a major challenge will be to connect the different levels of gene expression systems though large-scale data integration, which should lead to new systems-level insights into the logic of cellular and physiological function (Brockmann *et al*, 2007; Lackner *et al*, 2007).

5. Emerging principles of the RNA-protein interaction network

To date, RIP-Chip or related protocols have been employed to more than 50 RBPs from yeast, some RBPs from worms, flies, plants, and over 20 human RBPs (for a comprehensive review on RIP/CLIP experiments see Morris *et al*, 2010). Besides specific insights into the cellular role of the particular RBPs under investigation, these studies strongly support and further extended the 'post-transcriptional operon model' initially proposed by Jack Keene and colleagues (Keene and Tenenbaum, 2002). In analogy to prokaryotic operons, this model predicts that RBPs in eukaryotes coordinate groups of mRNAs coding for functionally related proteins. *Cis-acting* elements in the mRNA may provide the means to mimic the coordinated regulatory advantages of clustering genes into polycistronic operons (Keene, 2007; Keene *et al*, 2002). In the following, I will briefly summarize the major insights from RIP-Chip experiments.

First, RBPs bind to unique sets of RNA. Thereby, the number of the associated RNAs can vary widely as shown in our ribonomic study of 46 yeast RBPs revealing unique associations with 20 to >1,000 distinct transcripts per RBP (Hogan et al, 2008). Second, sequence or structural elements in the RNAs are enriched among targets often defining the binding site for the RBP. Using bioinformatic tools, we have identified diverse sequence/structural elements among the bound RNAs and verified that they are sufficient to direct protein interaction in several instances (Galgano et al, 2008; Gerber et al, 2004; Gerber et al, 2006; Hasegawa et al, 2008; Hogan et al, 2008; Scherrer et al, 2011). Third, bound mRNAs often encode functionally and/or cytotopically related proteins. This is perhaps best exemplified with the yeast Puf proteins, the targets of which share striking common cytotopic features (Gerber et al, 2004). Fourth, it appears that RBPs tend to regulate other regulatory proteins such as RBPs and transcription factors. We further propose that post-transcriptional interactions between RBPs form a dense and intertwined network, with potential for nearly half of the RBPs to auto-regulate transcript levels. As such, there is higher incidence for potential autoregulatory loops among RBPs than among TFs (10%) (Kanitz and Gerber, 2009) (Fig. 8). These observations suggest that post-transcriptional gene regulation (PTGR) is rather robust, self-sustaining and causes changes in gene expression independent of a transcriptional input event (e.g. our translatome analysis of cells exposed to different sorts of stress; Appendix II, Halbeisen et al, 2009a). Fifth, the spectra of targets of RBPs strongly overlap with targets of other RBPs, suggesting strong combinatorial binding of RBPs. Such combinatorial regulation may greatly impinge on the regulatory potential of the RBPs - breaking-up simple linear correlations between RBPs and the fates of its targets. For instance, we have examined this for Khd1, a yeast protein that oppositely controls the expression of mRNA targets: some messages are increasingly expressed ("up-regulated") whereas others are repressed ("downregulated") by Khd1p (Hasegawa et al, 2008).

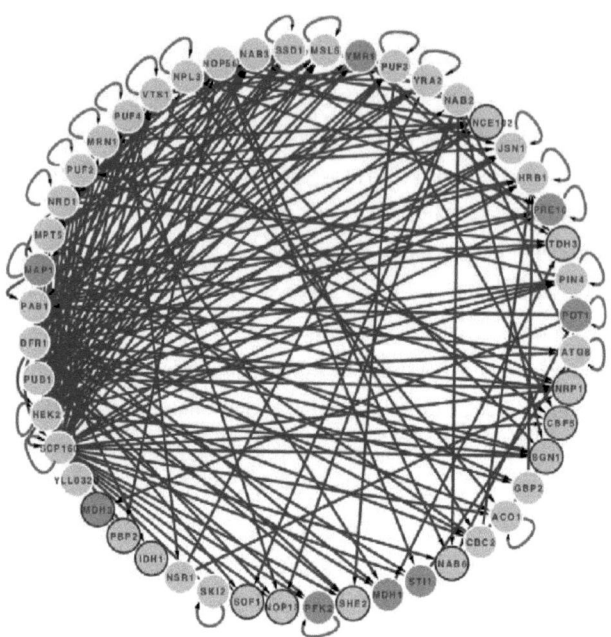

Fig. 8 Dense network of post-transcriptional regulatory interactions between yeast RBPs.

To explore the cross-talk and post-transcriptional regulatory potential between RBPs, we analyzed the posttranscriptional regulatory network comprised of RBPs as potential regulators, and their corresponding mRNAs as targets. We retrieved RNA-protein interaction data for 40 canonical RBPs from (Hogan *et al*, 2008) and data for eight "novel" RBPs using protein microarrays are shown in green (Scherrer T, Gerber AP, submitted). The network reveals 36 RBPs as regulators and 48 RBPs as targets, adding up to 221 mRNA-protein interactions. The circular layout was created with Cytoscape (Cline *et al*, 2007), where each node corresponds to one RBP with connections between them corresponding to protein-mRNA interactions shown in blue. RBPs are sorted clock-wise with decreasing connections to other RBPs. Associations of RBPs with their own mRNA are shown with loops colored in red. RBPs framed in brown are exclusively targets of other RBPs (Network structure was created in collaboration with Dr. Sarath J. Chandra, Medical Research Council, Cambridge, UK).

Sixth, the RNA-protein network appears to be very dynamic and responds to environmental or developmental signals by altering the RNA and protein content of RNPs. Thereby, post-translational modification of the RBP (e.g. phosphorylation) can alter subcellular localization or RNA-binding activity of the RBP (Huttelmaier *et al*, 2005; Paquin *et al*, 2007). For instance, we have observed different sets of mRNAs bound to PUMILIO proteins in embryos and adult flies, which cannot be solely explained by the altered gene expression at the two developmental stages

(Gerber et al, 2006). The dense network of interconnections between RBPs could also enable to respond quickly to different cues and control the RNA levels by the combinatorial interplay of RBPs, which form dynamic RNPs (Fig. 8).

Finally, I wish to add two evolutionary aspects: On one hand, we have obtained data that distinct yet overlapping groups of transcripts can be coordinated by paralogous RBPs. Besides our analysis of the paralogous human PUM proteins (Galgano et al, 2008), the determination of the RNA targets for the two yeast TRAMP complexes (TRAMP4/5) suggest overlapping yet distinct roles for decay of RNAs (San Paolo et al, 2009). Moreover, we have evidence for the evolvement of new mRNA regulatory circuits by paralogous RNPs that may be used for cellular adaptation to specific environmental conditions (Luca Schenk and APG, unpublished results). On the other hand, as we have shown for PUF-family proteins, the protein domains and the sequence motifs responsible for protein binding may be well conserved during evolution but the identities of the proteins and the mRNAs, which contain these proteins and motifs do not have to be necessarily conserved as well. We have therefore speculated that this discordance suggests that acquisition or loss of RBP-binding motifs in UTRs of genes may provide a surprisingly fluid evolutionary mechanism to modify post-transcriptional regulatory connections (Gerber et al, 2006). On the other hand, we have lately found strong structural and functional conservation among mRNA targets for conserved ZnF RBPs from yeast and human (Scherrer et al, 2011). In this case however, the proteins preferentially bind to sequence motifs located in coding sequences of mRNA targets, which have been generally better conserved than UTR sequences during evolution.

RPBs tend to 'coordinate' groups of functionally related messages. In this regard, 'coordination' is different from 'control' - while 'control' describes an individual RBP-RNA interaction that results in a specific outcome, 'coordination' describes a process of integrating multiple control functions to achieve a higher level of harmonization in the outcome (Mesarovic et al, 2004). The phenotypic consequences of coordination may reflect the combined effects of the trans-acting RBP (or miRNA) on multiple targets. The RNA-protein network is further complicated by the possibility that multiple states of a specific mRNA population may exist and therefore, not every RNA molecule may have the same fate. For example, if there are 10 copies of a mRNA x in the cell, 5 may be bound by RBP a, and actively translated in the cytoplasm, 3 may be bound by RBP b and stored in an inactive state, and 2 may be transported to a subcellular compartment by RBP c. Therefore, the steady-state levels of specific mRNAs does not have to alter upon changing conditions (e.g. tumorigenesis, cell differentiation, stress), but the organization of the RBP-protein interaction network (=ribonome) may considerably change with implications for translation or the subcellular location of messages.

6. RNP networks in human disease

Considering that RBPs are key cellular components coordinating functionally related groups of messages, it is evident that defects in their function should be commonly observed in human disease as their RNA-binding capacity can impact many different genes and pathways For instance, alteration of post-transcriptional gene regulation can mark disease such as alternative splicing of mRNAs in cancer (Tazi *et al*, 2009), or can even directly cause disease as exemplified by nucleotide repeat expansions in neurodegenerative disorders (Cooper *et al*, 2009). In the following, I will review some aspects of the roles for RNA and RBPs in disease. I also briefly discuss some of our own investigations on Gis2/ZNF9 – conserved ZnF proteins from yeast and human for which we identified the RNA targets. Because ZNF9 is hereditary linked to myotonic dystrophy (DM), we thereby got some unexpected clues raising speculations about potential implications of ZNF9 in this disease.

6.1 RBP loss of function

To date, about 30 hereditary diseases have been linked to mutation in genes coding for RBPs (Table 2) (reviewed in Cooper *et al*, 2009; Lukong *et al*, 2008). Intriguingly, it appears that many RBP defects are linked to muscular and neurodegenerative disorders. The reason for this preference is not known but it indicates that post-transcriptional regulation is highly regulated in these tissues and cell-types. An increasing number of RBPs has also become linked to cancer as they can act as oncogenes or tumor suppressors (Galante *et al*, 2009; Kim *et al*, 2009); and they coordinate the initiation and resolution of inflammation (Anderson, 2010).

Table 2 Disease implications of RBPs and RNA metabolism.

Disease	Abbrev.	Gene/Mutation	Function
RBP loss-of-function			
Lupus erythematosus, Sjogren syndrome	SLE	LARP3	transcription PolIII, tRNA maturation
Amyotrophic lateral sclerosis	ALS	TARDBP	transcription, splicing, mRNA stability
Dyskeratosis congenital (X-linked)	DKC	DKC1	Defects in RNP telomerase activity, translation
Dyskeratosis congenital (autosomal)	DKC	TERC, TERT	Defects in RNP telomerase activity
familial dysautonomia	FD	IKAP	elongator complex, tRNA
(Non)Syndromic mental retardation	(N)MRS	UPF3B	Nonsense-mediated decay
Cancer		SFRS1	alternative splicing
Cancer		RBM5	alternative splicing
autism, mental retardation, epilepsy		A2BP1/FOX1A	alternative splicing
Opsoclonus-myoclonus ataxia	POMA	NOVA 1,2,3	Alternative splicing
Retinitis pigmentosa	RP	PRPF32, PRPF8, HPF	SnRNP assembly
Spinal muscular atrophy	SMA	SMN1, SMN2	RNP assembly, RNP localization
Diamond-Blackfan anemia	DBA	RPS19, RPS24	Ribosome biogenesis 40S subunit
Shwachman-Bodian-Diamond syndrome	SDS	SBDS	Ribosome biogenesis of 60S subunit
Treacher Collins-Franceschetti syndrome 1		TCOF1	Ribosomal DNA transcription, RNA transport
Oculopharyngeal muscular dystrophy	OPMD	PABPN1	Polyadenylation
Dyschromatosis symmetrica hereditaria	DHS	ADAR1	Editing
Fragile X-syndrome	FKS	FMR1	Translation and RNA localization
Charcot-Marie-Tooth neuropathy 2D	CMT2D	GARS	translation
Charcot-Marie-Tooth	CMT	YARS	translation
Leukoencephalopathy with vanishing white matter		EIF2B1,2,3,4,5	translation
Mitochondrial myopathy and sideroblastic anemia	MLASA	PUS1	translation, pseudouridine formation in tRNAs
Combined oxidative phosphorylation deficiency-3 syndrome	COXPD3	TSFM	translation, oxidative phosphorylation
Leukoencephalopathy	LBSL	DARS2	translation, mitochondria
Diabetes mellitus (?)		LARS2	translation, mitochondria
Schizophrenia	SCZ	QKI	translation, mRNA stability
Cancer		SAM68	translation, mRNA stability
Hu syndrome, cancer		ELAV	mRNA stability and export
RNA gain-of-function			
Muscular dystrophy type 1	DM1	DMPK	Alternative splicing; toxic RNA; RBP entrapment
Muscular dystrophy type 2	DM2	ZNF9	Alternative splicing; toxic RNA; RBP entrapment
Fragile-X-associated tremor/ataxia syndrome	FXTAS	FMR1	Toxic mRNA; RBP entrapment
Huntington disease-like 2	HDL-2	JPH3	RBP entrapment
Spinocerebellar ataxias	SCA	ATXN8, ATXN8OS	Splicing

neurological disease
muscular atrophies
Cancer

Perhaps the best known example of a disease caused by mutation in a RBPs genes refers to the Fragile-X-mental retardation protein (FMRP) causing Fragile X syndrome (FSX) (reviewed in Oostra and Willemsen, 2009). FSX is the most common inherited form of mental disorder (incidence in males 1:4000), affecting higher-cognitive functions and leading to craniofacial anomalies. The syndrome is caused by a CGG triplet expansion of >200 repeats located within the 5'-UTR of the *FMR1* gene. In healthy individuals, the number of CGG repeats is variable and extends between 6 and 54. Hypermethylation of the CpG repeats leads to gene silencing and thus, the corresponding protein FMRP is absent in FXS patients. FMRP harbors two KH domains and one RGG-type RNA-binding domain and binds to RNA with high affinity. In neurons, FMRP is present in cytoplasmic RNPs associated with somatic and synaptic polyribosomes, and is thought to repress the translation of specific messages during their transport to dendrites and at the synapse, which is now recognized to be an important process for modulation of synaptic plasticity and memory formation. However, how FMRP positively or negatively regulates translation is still unclear - possibly FMRP mediates repression through the RNA interference pathway, interacting with components of the RISC complex. It is has also been hypothesized that FMRP controls the translation of mRNAs encoding proteins that regulate endocytic events of AMPA receptors at the synapse (AMPA receptors are a type of glutamate receptors; glutamate being the major excitatory neurotransmitter in the central nervous system). Upon synaptic stimulation, FMRP dissociates from these mRNA allowing their translation, which results in the uptake of AMPA receptors and long-term depression (LTD). In this model, absence of FMRP would disrupt this uptake and lead to perturbation of AMPA receptor internalization. Following this model, new treatments of FXS focus on metabotropic glutamate receptors antagonists that can control the uptake of AMPA receptors in neurons. After successful application in mouse models of FXS, Phase I clinical trials have now been undertaken with Fenobam and AFQO56, and no side-effects have been reported so far (Oostra *et al*, 2009).

Other examples where mutations in RBPs lead to disease include mutations in the Nova genes causing paraneoplastic opsoclonus-myoclonus ataxia (POMA), and SMN1 or SF2 in spinal muscular atrophy (SMA) (Table 2). These RBPs are splicing regulators (Nova, SF2) and their mutation may lead to mis-splicing, particularly in neurons leading to a variety of neurological phenotypes and disease (Cooper *et al*, 2009; Lukong *et al*, 2008). As the interest in RBP-mediated post-transcriptional gene regulation is rapidly increasing, there will certainly be new discoveries that connect misregulation or absence RBPs to disease.

Many RBPs are also differentially expressed in tumor models (Galante *et al*, 2009) – however, generally it is not clear whether the elevated expression of RBPs in certain tumors is a cause or a consequence of cancer (Kim *et al*, 2009; Tazi *et al*, 2009). In fact, there are only a few RBPs identified to directly act as oncogenes (e.g. CRD-BP, RBM3, PTB, Musashi 1, and ASF/SF2) or as tumor suppressors (RBM5, Luca-12) (Galante *et al*, 2009). For most of these proteins the RNA targets have not been systematically explored yet and thus, ribonomic profiling of disease-related RBPs is in need to unravel the pathways regulated by them. The knowledge of these networks may further allow the development of novel approaches for diagnosis and treatment of cancer.

6.2 Mutations in *cis*-acting RNA elements

Next to mutations of *trans*-acting factors such as RBPs and miRNAs (Mencia *et al*, 2009), single point-mutations in *cis*-acting elements that define RBP/miRNA binding sites can also cause disease (Chatterjee and Pal, 2009; Cooper *et al*, 2009). Most prominent are specific mutations that cause splicing defects. Splice sites are marked by short sequence elements that define introns in pre-mRNAs. In addition, a variety of so-called splice-site enhancers or repressors exist that recruit protein factors to modulate splicing (e.g. Nova). The mutation of either of these sites can disrupt interactions with splicing factors and hence, alters splicing possibly leading to disease. Remarkably, it is thought that such mutations are responsible for about 10% of the genetic diseases caused by point mutations (Cooper *et al*, 2009).

Although gene expression regulation via UTRs is now increasingly recognized, most pharmacogenetics studies still focus on the coding sequences leaving UTRs largely unconsidered. Therefore, careful re-evaluation of single-nucleotide polymorphism (SNP) or copy number variants (CNVs) in UTRs may reveal novel, previously unrecognized instances for disease, which may lead to new markers for diagnosis. For instance, several recent studies have found 3'UTR SNPs that affect gene expression via miRNA gene regulation in different diseases. In these cases, the disruption of miRNA targets binding sites by SNPs or other mutations of sequences in UTRs has been associated with disease i.e. cancer susceptibility and initiation (reviewed in Chen *et al*, 2008). Examples include a kRAS variant (a component of an essential signaling pathway) within let-7 miRNA target sites, which increases the risk for non-small cell lung carcinoma among moderate smokers (Chin *et al*, 2008). Polymorphisms of a *mir-24* binding site in the dihydrofolate reductase gene leads to methotrexate resistance (Mishra *et al*, 2007), and a recent study reported that SNPs inside miRNA targets sites are correlated with the pathogenetic relevance of known breast cancer associated SNPs and cancer susceptibility (Nicoloso *et al*, 2010). In this regard, the comprehensive analysis of the frequencies of miRNA-binding SNPs in cancers versus normal tissues through the mining of EST databases or other sources found dozens of SNPs that are potentially associated with disease/ cancer, indicating that miRNA deregulation by SNPs may be a rather widespread mechanisms that influences cancer susceptibility (Landi *et al*, 2008; Yu *et al*, 2007). Intriguing, the analogous analysis of frequencies of SNPs in binding sites for RBPs has not been conducted yet. Since RBPs impinge on all major physiological processes, I am sure that such an analysis may provide new and unexpected findings.

6.3 RNA gain-of-function and myotonic dystrophy

Besides the very specific and minimal mutations in *cis*-acting elements, longer repeat expansions, mostly located in introns or UTRs, can cause the production of RNA structures that could become toxic to cells. RNA gain-of-function is achieved when such RNA structures capture specific RPBs, thereby forming intracellular aggregates withdrawing the captured RBPs from their normal default targets (Table 2). One well described example for this RNA gain-of-function mechanism relates to muscular dystrophy (DM), a multi-systemic neuromuscular disease characterized by heterogeneous, multi systemic symptoms including myotonia, progressive muscle weakness and wasting, cataracts, and cardiac conduction defects (incidence of about 1:4000 people) (Schara and Schoser, 2006). Two types of DM have been described: CTG expansion in the 3'UTR of DMPK cause DM1 and CCTG repeat expansions in the first intron of ZNF9 cause DM2 (Ranum and Cooper, 2006). In the case of DM2, the first intron of *ZNF9* bearing CCUG expansions is apparently correctly spliced, but accumulates in nuclear foci sequestering members of the muscleblind-like (MBNL) family of RBPs (Cho and Tapscott, 2007; Mankodi, 2008; Ranum *et al*, 2006). The depletion of MBNL1 and other RBPs by the RNA repeats leads then to the abnormal splicing of chloride channel 1 *CLCN1* and troponin *TNNT2* mRNAs and thus, the phenotypic consequences of ZNF9 intron repeat expansions are primarily thought to be indirect and disregard a direct role for ZNF9 in pathogenesis. This idea was substantiated by studies that reported no alterations of ZNF9 mRNA or protein levels in DM2 patients compared to healthy individuals (Margolis *et al*, 2006). However, other and more recent evidence also support a direct role of ZNF9 in DM2, such as $ZNF9^{+/-}$ heterozygous mice, which develop symptoms related to DM2 (Chen *et al*, 2007); and reduced levels of ZNF9 in myoblasts of DM2 patients compared to healthy individuals have also been seen (Huichalaf *et al*, 2009). Therefore, it is still not ruled-out whether or not ZNF9 may also have a direct role in the pathogenesis of DM2.

Intriguingly, ZNF9 is a nucleic acid binding protein that bears seven Zinc-finger (ZnF) domains and one RGG-box domain, both of which could mediate RNA or DNA-protein interactions (see also Table 1). ZNF9, also referred to as cellular nucleic acid binding protein (CNBP), was first described to bind to purine-rich single-stranded (ss)DNA of the sterol response element (SRE) possibly playing a role in sterol metabolism (Rajavashisth *et al*, 1989). Several studies further suggested functions of ZNF9 as a positive or negative regulator of transcription by binding to guanosine-rich ssDNA sequences. ZNF9 is highly conserved across the seven ZnFs in eukaryotes, which includes related proteins in the budding yeast *Saccharomyces cerevisiae* (Gis2), flies (*Drosophila melanogaster* [CG3800]) and worms *Caenorhabditis elegans* [K08D12.3]. Nevertheless, the RGG box is only present in vertebrates. In conclusion, ZNF9 and homologous proteins in other species may directly modulate gene expression, either by binding to DNA promoter elements upstream of genes that they activate/repress, or post-transcriptionally through binding to RNAs.

In order to investigate the roles of ZNF9 and related protein for RNA regulation, we have therefore applied an integrated genomic and proteomic approach to explore the role of this protein in RNA expression regulation (Scherrer *et al*, 2011). We thereby focused on the related protein from yeast, termed Gis2p, for which we first identified bound RNAs in cells applying RIP-Chip. Almost 1,000 different mRNAs were associated with Gis2 *in vivo*, mainly coding for RNA processing factors, chromatin modifiers, and GTPases – reminiscent to post-transcriptional operons. A matched-sample proteome-transcriptome analysis revealed that Gis2 could differentially balance the expression of functionally related mRNA targets – namely mRNAs that code for proteins acting in rRNA processing and act in the nucleolus. We also found a striking functional conservation between the yeast and the human RBP, suggesting yeast as a suitable model for studying functions of *ZNF9*: Target mRNAs contained stretches of G(A/U)(A/U) trinucleotide repeats, which were primarily located in coding sequences. We further showed with RNA pull-down experiments that these repeats are sufficient for binding to both Gis2 and human ZNF9, thus implying strong structural conservation between the orthologous proteins. Based on our definition of the *cis*-acting RNA recognition elements for Gis2 and ZNF9, we could predict the putative targets for human ZNF9 with bioinformatics tools. Strikingly, these predicted ZNF9 targets were enriched for the same functional categories as seen in yeast, indicating also strong functional conservation (Fig. 9). The strong functional relation was further supported by the complementation of the large cell-size phenotype of *gis2* mutants with *ZNF9*. Additionally, we found that ZNF9 could target many mRNAs coding for proteins involved in muscle contraction such as myosins and ion channels and hence, ZNF9 may coordinate expression of a variety of genes with pivotal roles in muscle function. Although this idea has to be substantiated, we speculate that altered expression of these genes may relate to reported phenotypes seen with heterozygous *ZNF9* knock-out mice, which resemble the abnormalities seen in DM2 patients.

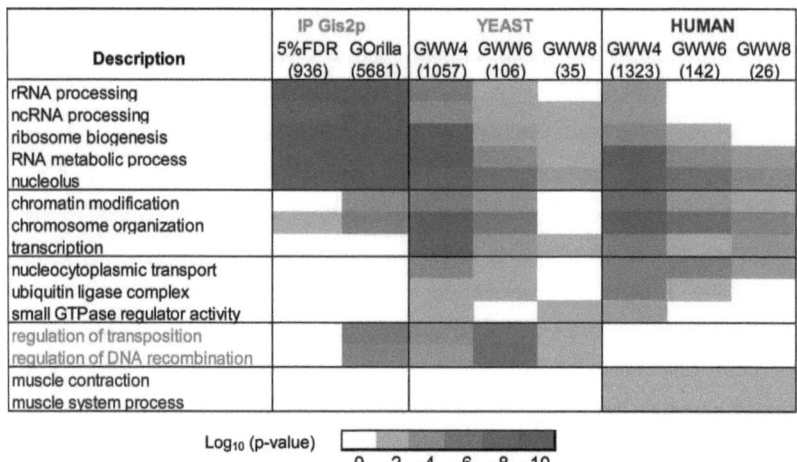

Fig. 9 Significantly shared GO terms among experimentally defined and predicted Gis2 and ZNF9 mRNA targets.

GO term enrichments were assessed for experimentally detected (RIP-Chip analysis of Gis2 targets) or predicted mRNA targets bearing different length of GWW repeats in coding sequences. Numbers in brackets indicate the total number of genes in the respective group for which GO annotations were available. The bw intensity corresponds to the $\log_{10}$ p-value.

6.4 Perspectives for drug development and therapy

Ribonomic analysis of RBPs and determination of the *cis*-acting elements in the transcripts to which they bind will greatly enhance our understanding of the mode of action of RBPs and ncRNAs, and possibly open new approaches to cure disease. For instance, the emerging relation between HuR and cancer has increased interest for HuR as a drug target. HuR is a member of the Hu (ELAV) family of RBPs that binds to AU-rich elements in 3'-UTRs of messages. The HuR binding to mRNAs is important for stabilization of relatively short-lived mRNA targets coding for proto-oncogenes, anti-apoptotic proteins, proangiogenic growth factors and proteins essential for cell migration. Recently, Novartis has been developing several low-molecular-weight inhibitors of HuR, which are now considered for further testing as novel anti-cancer reagents (Meisner *et al*, 2007). The compounds were identified by high-throughput screening of microbial broths (*Actinomyces sp.*), selecting several compounds (dehydromutactin, MS-444 and okicenone) that interfere with HuR RNA binding, HuR trafficking, cytokine expression and T-cell activation. These results demonstrated the chemical drugability of HuR and these compounds could become valuable tools for studying HuR function. Assessment of HuR inhibition as a central node in malignant processes might also open new conceptual routes toward combating cancer (Meisner *et al*, 2007). Another approach was taken by the Keene lab: the comparison of the Connectivity Map (also known as cmap; a set of gene expression profiles obtained from cells treated with a panel of bioactive small molecules) with profiles of HuR-associated transcripts in activated T cells identified compounds that modulate HuR-mediated post-transcriptional control (Mukherjee *et al.* 2009). This latter approach could be more broadly used to identify molecules that regulate post-transcriptional control by RBPs that target the 3'-UTR or other regions of selected transcripts. Small molecules are also attractive for pharmacological modulation of alternative splicing. As the splicing of most introns is strongly dependent on serine-arginine proteins and hnRNP proteins, the development of small molecules targeting their activity has great potential to modify splicing and possibly to cure disease. Indeed, several compounds targeting SR proteins have been developed that hold promise to interfere with splicing events *e.g.* HIV-1 splicing (Cooper *et al*, 2009).

Strong efforts are also undertaken to develop tools to the correct for mis-splicing with antisense RNA technologies (ASOs) (Bonetta, 2009). Generally, several AOs have been designed or are under development to hybridize and block sequences in the targets pre-mRNA that are critical for a particular splicing event. Examples of disease-causing genes currently targeted by AOs include the β-globin gene in β-thalassemia, the CFTR gene in cystic fibrosis; and AOs represent potential treatment strategies for SMA and to correct splicing defects in Duchenne muscular dystrophy (DMD). Future will show whether these new drug targets are successful and lead to new therapties for the treatment of human disease.

The first RNA-based drug for the treatment of a human disease was approved in 1998 by the US Food and Drug Administration (FDA). The drug, called fomivirsen, is an AO that blocks synthesis of a key protein of cytomegalovirus and is used to treat inflammation of the eye caused by the virus. To date, although only one other RNA-based drug (an RNA aptamer) has made it to market, the arsenal of potential molecules and approaches continues to expand. It now includes a new generation of antisense oligos, aptamers, ribozymes, RNA decoys, splice-site targeted oligos, small-interfering RNAs (siRNAs), short hairpin RNAs (shRNAs), and microRNAs (miRNAs). Although they all target RNA, these compounds work by distinct mechanisms. However, they share similar hurdles to clinical application— the biggest of which is efficient delivery to the desired tissue. In conclusion, the design and application of these new therapeutic approaches are promising – however, there will certainly be many hurdles to be taken before this new generation of compounds that modulate post-transcriptional gene expression will appear on the market.

7. Concluding Remarks

The advent of global and quantitative analysis tools for the study of gene expression allowed the detection and quantification of network motifs in gene regulatory systems. Generally, it appears that many of the principles and structures of transcriptional regulatory networks are also preserved at the post-transcriptional level. However, systems analysis of post-transcriptional gene regulation is still in its infancy. The development of novel techniques for RNA network analysis will hence be crucial to obtain sufficient data for understanding the "RNP code". The combination of ribonomic approaches with crosslinking techniques and high-throughput sequencing will help to systematically map RBP binding sites. The use of next generation sequencing methods will allow a more robust and quantitative detection of RNAs, including rare and unknown RNA molecules/species. Furthermore, the application of quantitative proteomics should permit the quantitative description of mRNA translation and degradation. The data obtained from such studies can then be used for the development/refinement of mathematical models of gene regulation that will improve in accuracy and predictive power. Such analyses will certainly lead to new systems-level insights into the logic of cellular and physiological functions and disease-causing perturbations and, hence, will hopefully lead to new and unexpected approaches for their cure.

8. References

Anantharaman V, Koonin EV, Aravind L (2002) Comparative genomics and evolution of proteins involved in RNA metabolism. *Nucleic Acids Res* 30: 1427-1464.

Anderson P (2010) Post-transcriptional regulons coordinate the initiation and resolution of inflammation. *Nat Rev Immunol* 10: 24-35.

Asaoka-Taguchi M, Yamada M, Nakamura A, Hanyu K, Kobayashi S (1999) Maternal Pumilio acts together with Nanos in germline development in Drosophila embryos. *Nat Cell Biol* 1: 431-437.

Auweter SD, Oberstrass FC, Allain FH (2006) Sequence-specific binding of single-stranded RNA: is there a code for recognition? *Nucleic Acids Res* 34: 4943-4959.

Baek D, Villen J, Shin C, Camargo FD, Gygi SP, Bartel DP (2008) The impact of microRNAs on protein output. *Nature* 455: 64-71.

Baroni TE, Chittur SV, George AD, Tenenbaum SA (2008) Advances in RIP-chip analysis : RNA-binding protein immunoprecipitation-microarray profiling. *Methods Mol Biol* 419: 93-108.

Bartel DP (2009) MicroRNAs: target recognition and regulatory functions. *Cell* 136: 215-233.

Beilharz TH, Preiss T (2004) Translational profiling: the genome-wide measure of the nascent proteome. *Brief Funct Genomic Proteomic* 3: 103-111.

Beitzinger M, Peters L, Zhu JY, Kremmer E, Meister G (2007) Identification of Human microRNA Targets From Isolated Argonaute Protein Complexes. *RNA Biol* 4.

Beltran M, Puig I, Pena C, Garcia JM, Alvarez AB, Pena R, Bonilla F, de Herreros AG (2008) A natural antisense transcript regulates Zeb2/Sip1 gene expression during Snail1-induced epithelial-mesenchymal transition. *Genes Dev* 22: 756-769.

Bernstein DS, Buter N, Stumpf C, Wickens M (2002a) Analyzing mRNA-protein complexes using a yeast three-hybrid system. *Methods* 26: 123-141.

Bernstein JA, Khodursky AB, Lin PH, Lin-Chao S, Cohen SN (2002b) Global analysis of mRNA decay and abundance in Escherichia coli at single-gene resolution using two-color fluorescent DNA microarrays. *Proc Natl Acad Sci USA* 99: 9697-9702.

Bonetta L (2009) RNA-based therapeutics: ready for delivery? *Cell* 136: 581-584.

Bramham CR, Wells DG (2007) Dendritic mRNA: transport, translation and function. *Nat Rev Neurosci* 8: 776-789.

Brockmann R, Beyer A, Heinisch JJ, Wilhelm T (2007) Posttranscriptional expression regulation: what determines translation rates? *PLoS Comput Biol* 3: e57.

Butter F, Scheibe M, Morl M, Mann M (2009) Unbiased RNA-protein interaction screen by quantitative proteomics. *Proc Natl Acad Sci U S A* 106: 10626-10631.

Chatterjee S, Pal JK (2009) Role of 5'- and 3'-untranslated regions of mRNAs in human diseases. *Biol Cell* 101: 251-262.

Chekulaeva M, Filipowicz W (2009) Mechanisms of miRNA-mediated post-transcriptional regulation in animal cells. *Curr Opin Cell Biol* 21: 452-460.

Chen K, Song F, Calin GA, Wei Q, Hao X, Zhang W (2008) Polymorphisms in microRNA targets: a gold mine for molecular epidemiology. *Carcinogenesis* 29: 1306-1311.

Chen W, Wang Y, Abe Y, Cheney L, Udd B, Li YP (2007) Haploinsuffciency for Znf9 in Znf9+/- mice is associated with multiorgan abnormalities resembling myotonic dystrophy. *J Mol Biol* 368: 8-17.

Chi SW, Zang JB, Mele A, Darnell RB (2009) Argonaute HITS-CLIP decodes microRNA-mRNA interaction maps. *Nature* 460: 479-486.

Chin LJ, Ratner E, Leng S, Zhai R, Nallur S, Babar I, Muller RU, Straka E, Su L, Burki EA, Crowell RE, Patel R, Kulkarni T, Homer R, Zelterman D, Kidd KK, Zhu Y, Christiani DC, Belinsky SA, Slack FJ, Weidhaas JB (2008) A SNP in a let-7 microRNA complementary site in the KRAS 3' untranslated region increases non-small cell lung cancer risk. *Cancer Res* 68: 8535-8540.

Cho DH, Tapscott SJ (2007) Myotonic dystrophy: emerging mechanisms for DM1 and DM2. *Biochim Biophys Acta* 1772: 195-204.

Cline MS, Smoot M, Cerami E, Kuchinsky A, Landys N, Workman C, Christmas R, Avila-Campilo I, Creech M, Gross B, Hanspers K, Isserlin R, Kelley R, Killcoyne S, Lotia S, Maere S, Morris J, Ono K, Pavlovic V, Pico AR, Vailaya A, Wang PL, Adler A, Conklin BR, Hood L, Kuiper M, Sander C, Schmulevich I, Schwikowski B, Warner GJ, Ideker T, Bader GD (2007) Integration of biological networks and gene expression data using Cytoscape. *Nat Protoc* 2: 2366-2382.

Cooper TA, Wan L, Dreyfuss G (2009) RNA and disease. *Cell* 136: 777-793.

Costa-Mattioli M, Sossin WS, Klann E, Sonenberg N (2009) Translational control of long-lasting synaptic plasticity and memory. *Neuron* 61: 10-26.

Dreyfuss G, Kim VN, Kataoka N (2002) Messenger-RNA-binding proteins and the messages they carry. *Nat Rev Mol Cell Biol* 3: 195-205.

Dubnau J, Chiang AS, Grady L, Barditch J, Gossweiler S, McNeil J, Smith P, Buldoc F, Scott R, Certa U, Broger C, Tully T (2003) The staufen/pumilio pathway is involved in Drosophila long-term memory. *Curr Biol* 13: 286-296.

Edwards TA, Pyle SE, Wharton RP, Aggarwal AK (2001) Structure of Pumilio reveals similarity between RNA and peptide binding motifs. *Cell* 105: 281-289.

Eliyahu E, Pnueli L, Melamed D, Scherrer T, Gerber AP, Pines O, Rapaport D, Arava Y (2010) Tom20 mediates localization of mRNAs to mitochondria in a translation-dependent manner. *Mol Cell Biol* 30: 284-294.

Elson SL, Noble SM, Solis NV, Filler SG, Johnson AD (2009) An RNA transport system in Candida albicans regulates hyphal morphology and invasive growth. *PLoS Genet* 5: e1000664.

Erson AE, Petty EM (2008) MicroRNAs in development and disease. *Clin Genet* 74: 296-306.

Filipowicz W, Bhattacharyya SN, Sonenberg N (2008) Mechanisms of post-transcriptional regulation by microRNAs: are the answers in sight? *Nat Rev Genet* 9: 102-114.

Forbes A, Lehmann R (1998) Nanos and Pumilio have critical roles in the development and function of Drosophila germline stem cells. *Development* 125: 679-690.

Fox S, Filichkin S, Mockler TC (2009) Applications of ultra-high-throughput sequencing. *Methods Mol Biol* 553: 79-108.

Gaillard H, Aguilera A (2008) A novel class of mRNA-containing cytoplasmic granules are produced in response to UV-irradiation. *Mol Biol Cell* 19: 4980-4992.

Galante PA, Sandhu D, de Sousa Abreu R, Gradassi M, Slager N, Vogel C, de Souza SJ, Penalva LO (2009) A comprehensive in silico expression analysis of RNA binding proteins in normal and tumor tissue: Identification of potential players in tumor formation. *RNA Biol* 6.

Galgano A, Forrer M, Jaskiewicz L, Kanitz A, Zavolan M, Gerber AP (2008) Comparative analysis of mRNA targets for human PUF-family proteins suggests extensive interaction with the miRNA regulatory system. *PLoS ONE* 3: e3164.

Garcia-Rodriguez LJ, Gay AC, Pon LA (2007) Puf3p, a Pumilio family RNA binding protein, localizes to mitochondria and regulates mitochondrial biogenesis and motility in budding yeast. *J Cell Biol* 176: 197-207.

Gebauer F, Hentze MW (2004) Molecular mechanisms of translational control. *Nat Rev Mol Cell Biol* 5: 827-835.

Gerber AP, Herschlag D, Brown PO (2004) Extensive Association of Functionally and Cytotopically Related mRNAs with Puf Family RNA-Binding Proteins in Yeast. *PLoS Biol* 2: E79.

Gerber AP, Keller W (2001) RNA editing by base deamination: more enzymes, more targets, new mysteries. *Trends Biochem Sci* 26: 376-384.

Gerber AP, Luschnig S, Krasnow MA, Brown PO, Herschlag D (2006) Genome-wide identification of mRNAs associated with the translational regulator PUMILIO in Drosophila melanogaster. *Proc Natl Acad Sci U S A* 103: 4487-4492.

Glisovic T, Bachorik JL, Yong J, Dreyfuss G (2008) RNA-binding proteins and post-transcriptional gene regulation. *FEBS Lett* 582: 1977-1986.

Goldstrohm AC, Hook BA, Seay DJ, Wickens M (2006) PUF proteins bind Pop2p to regulate messenger RNAs. *Nat Struct Mol Biol* 13: 533-539.

Goldstrohm AC, Seay DJ, Hook BA, Wickens M (Wa2007) PUF protein-mediated deadenylation is catalyzed by Ccr4p. *J Biol Chem* 282: 109-114.

Grigull J, Mnaimneh S, Pootoolal J, Robinson MD, Hughes TR (2004) Genome-wide analysis of mRNA stability using transcription inhibitors and microarrays reveals posttranscriptional control of ribosome biogenesis factors. *Mol Cell Biol* 24: 5534-5547.

Grimson A, Farh KK, Johnston WK, Garrett-Engele P, Lim LP, Bartel DP (2007) MicroRNA targeting specificity in mammals: determinants beyond seed pairing. *Mol Cell* 27: 91-105.

Gutierrez RA, Ewing RM, Cherry JM, Green PJ (2002) Identification of unstable transcripts in Arabidopsis by cDNA microarray analysis: rapid decay is associated with a group of touch- and specific clock-controlled genes. *Proc Natl Acad Sci USA* 99: 11513-11518.

Gygi SP, Rochon Y, Franza BR, Aebersold R (1999) Correlation between protein and mRNA abundance in yeast. *Mol Cell Biol* 19: 1720-1730.

Hafner M, Landthaler M, Burger L, Khorshid M, Hausser J, Berninger P, Rothballer A, Ascano M, Jr., Jungkamp AC, Munschauer M, Ulrich A, Wardle GS, Dewell S, Zavolan M, Tuschl T (2010) Transcriptome-wide identification of RNA-binding protein and microRNA target sites by PAR-CLIP. *Cell* 141: 129-141.

Halbeisen RE, Galgano A, Scherrer T, Gerber AP (2008) Post-transcriptional gene regulation: from genome-wide studies to principles. *Cell Mol Life Sci* 65: 798-813.

Halbeisen RE, Gerber AP (2009a) Stress-Dependent Coordination of Transcriptome and Translatome in Yeast. *PLoS Biol* 7: e105.

Halbeisen RE, Scherrer T, Gerber AP (2009b) Affinity purification of ribosomes to access the translatome. *Methods* 48: 306-310.

Hartmuth K, Vornlocher HP, Luhrmann R (2004) Tobramycin affinity tag purification of spliceosomes. *Methods Mol Biol* 257: 47-64.

Hasegawa Y, Irie K, Gerber AP (2008) Distinct roles for Khd1p in the localization and expression of bud-localized mRNAs in yeast. *RNA* 14: 2333-2347.

Heiman M, Schaefer A, Gong S, Peterson JD, Day M, Ramsey KE, Suarez-Farinas M, Schwarz C, Stephan DA, Surmeier DJ, Greengard P, Heintz N (2008) A translational profiling approach for the molecular characterization of CNS cell types. *Cell* 135: 738-748.

Hendrickson DG, Hogan DJ, Herschlag D, Ferrell JE, Brown PO (2008) Systematic identification of mRNAs recruited to argonaute 2 by specific microRNAs and corresponding changes in transcript abundance. *PLoS ONE* 3: e2126.

Hendrickson DG, Hogan DJ, McCullough HL, Myers JW, Herschlag D, Ferrell JE, Brown PO (2009) Concordant regulation of translation and mRNA abundance for hundreds of targets of a human microRNA. *PLoS Biol* 7: e1000238.

Hereford LM, Rosbash M (1977) Number and distribution of polyadenylated RNA sequences in yeast. *Cell* 10: 453-462.

Hieronymus H, Silver PA (2003) Genome-wide analysis of RNA-protein interactions illustrates specificity of the mRNA export machinery. *Nat Genet* 33: 155-161.

Hieronymus H, Silver PA (2004) A systems view of mRNP biology. *Genes Dev* 18: 2845-2860.

Hogan DJ, Riordan DP, Gerber AP, Herschlag D, Brown PO (2008) Diverse RNA-binding proteins interact with functionally related sets of RNAs, suggesting an extensive regulatory system. *PLoS Biol* 6: e255.

Hogg JR, Collins K (2007) RNA-based affinity purification reveals 7SK RNPs with distinct composition and regulation. *RNA* 13: 868-880.

Hollien J, Weissman JS (2006) Decay of endoplasmic reticulum-localized mRNAs during the unfolded protein response. *Science* 313: 104-107.

Huang YS, Richter JD (2004) Regulation of local mRNA translation. *Curr Opin Cell Biol* 16: 308-313.

Huichalaf C, Schoser B, Schneider-Gold C, Jin B, Sarkar P, Timchenko L (2009) Reduction of the rate of protein translation in patients with myotonic dystrophy 2. *J Neurosci* 29: 9042-9049.

Huttelmaier S, Zenklusen D, Lederer M, Dictenberg J, Lorenz M, Meng X, Bassell GJ, Condeelis J, Singer RH (2005) Spatial regulation of beta-actin translation by Src-dependent phosphorylation of ZBP1. *Nature* 438: 512-515.

Inada M, Guthrie C (2004) Identification of Lhp1p-associated RNAs by microarray analysis in Saccharomyces cerevisiae reveals association with coding and noncoding RNAs. *Proc Natl Acad Sci USA* 101: 434-439.

Jensen KB, Darnell RB (2008) CLIP: crosslinking and immunoprecipitation of in vivo RNA targets of RNA-binding proteins. *Methods Mol Biol* 488: 85-98.

Kadyrova LY, Habara Y, Lee TH, Wharton RP (2007) Translational control of maternal Cyclin B mRNA by Nanos in the Drosophila germline. *Development* 134: 1519-1527.

Kanitz A, Gerber AP (2009) Circuitry of mRNA regulation. *WIRES Syst Biol Med* 2: wsbm.55.

Karginov FV, Conaco C, Xuan Z, Schmidt BH, Parker JS, Mandel G, Hannon GJ (2007) A biochemical approach to identifying microRNA targets. *Proc Natl Acad Sci U S A*.

Keene JD (2007) RNA regulons: coordination of post-transcriptional events. *Nat Rev Genet* 8: 533-543.

Keene JD, Tenenbaum SA (2002) Eukaryotic mRNPs may represent posttranscriptional operons. *Mol Cell* 9: 1161-1167.

Kim MY, Hur J, Jeong S (2009) Emerging roles of RNA and RNA-binding protein network in cancer cells. *BMB Rep* 42: 125-130.

Kuersten S, Goodwin EB (2003) The power of the 3' UTR: translational control and development. *Nat Rev Genet* 4: 626-637.

Kunitomo H, Uesugi H, Kohara Y, Iino Y (2005) Identification of ciliated sensory neuron-expressed genes in Caenorhabditis elegans using targeted pull-down of poly(A) tails. *Genome Biol* 6: R17.

Lackner DH, Beilharz TH, Marguerat S, Mata J, Watt S, Schubert F, Preiss T, Bahler J (2007) A network of multiple regulatory layers shapes gene expression in fission yeast. *Mol Cell* 26: 145-155.

Landi D, Gemignani F, Barale R, Landi S (2008) A catalog of polymorphisms falling in microRNA-binding regions of cancer genes. *DNA Cell Biol* 27: 35-43

Landthaler M, Gaidatzis D, Rothballer A, Chen PY, Soll SJ, Dinic L, Ojo T, Hafner M, Zavolan M, Tuschl T (2008) Molecular characterization of human Argonaute-containing ribonucleoprotein complexes and their bound target mRNAs. *RNA*.

Lecuyer E, Yoshida H, Parthasarathy N, Alm C, Babak T, Cerovina T, Hughes TR, Tomancak P, Krause HM (2007) Global analysis of mRNA localization reveals a prominent role in organizing cellular architecture and function. *Cell* 131: 174-187.

Licatalosi DD, Mele A, Fak JJ, Ule J, Kayikci M, Chi SW, Clark TA, Schweitzer AC, Blume JE, Wang X, Darnell JC, Darnell RB (2008) HITS-CLIP yields genome-wide insights into brain alternative RNA processing. *Nature* 456: 464-469.

Lin H, Spradling AC (1997) A novel group of pumilio mutations affects the asymmetric division of germline stem cells in the Drosophila ovary. *Development* 124: 2463-2476.

Lopez de Silanes I, Fan J, Galban CJ, Spencer RG, Becker KG, Gorospe M (2004) Global analysis of HuR-regulated gene expression in colon cancer systems of reducing complexity. *Gene Expr* 12: 49-59.

Lopez de Silanes I, Galban S, Martindale JL, Yang X, Mazan-Mamczarz K, Indig FE, Falco G, Zhan M, Gorospe M (2005) Identification and functional outcome of mRNAs associated with RNA-binding protein TIA-1. *Mol Cell Biol* 25: 9520-9531.

Lu P, Vogel C, Wang R, Yao X, Marcotte EM (2007) Absolute protein expression profiling estimates the relative contributions of transcriptional and translational regulation. *Nat Biotechnol* 25: 117-124.

Lukong KE, Chang KW, Khandjian EW, Richard S (2008) RNA-binding proteins in human genetic disease. *Trends Genet* 24: 416-425.

Lunde BM, Moore C, Varani G (2007) RNA-binding proteins: modular design for efficient function. *Nat Rev Mol Cell Biol* 8: 479-490.

Macdonald PM (1992) The Drosophila pumilio gene: an unusually long transcription unit and an unusual protein. *Development* 114: 221-232.

Maniatis T, Reed R (2002) An extensive network of coupling among gene expression machines. *Nature* 416: 499-506.

Mankodi A (2008) Myotonic disorders. *Neurol India* 56: 298-304.

Margolis JM, Schoser BG, Moseley ML, Day JW, Ranum LP (2006) DM2 intronic expansions: evidence for CCUG accumulation without flanking sequence or effects on ZNF9 mRNA processing or protein expression. *Hum Mol Genet* 15: 1808-1815.

Mata J, Marguerat S, Bahler J (2005) Post-transcriptional control of gene expression: a genome-wide perspective. *Trends Biochem Sci* 30: 506-514.

Matsumoto M, Setou M, Inokuchi K (2007) Transcriptome analysis reveals the population of dendritic RNAs and their redistribution by neural activity. *Neurosci Res* 57: 411-423.

Meisner NC, Hintersteiner M, Mueller K, Bauer R, Seifert JM, Naegeli HU, Ottl J, Oberer L, Guenat C, Moss S, Harrer N, Woisetschlaeger M, Buehler C, Uhl V, Auer M (2007) Identification and mechanistic characterization of low-molecular-weight inhibitors for HuR. *Nat Chem Biol* 3: 508-515.

Mencia A, Modamio-Hoybjor S, Redshaw N, Morin M, Mayo-Merino F, Olavarrieta L, Aguirre LA, del Castillo I, Steel KP, Dalmay T, Moreno F, Moreno-Pelayo MA (2009) Mutations in the seed region of human miR-96 are responsible for nonsyndromic progressive hearing loss. *Nat Genet* 41: 609-613.

Menon KP, Sanyal S, Habara Y, Sanchez R, Wharton RP, Ramaswami M, Zinn K (2004) The translational repressor Pumilio regulates presynaptic morphology and controls postsynaptic accumulation of translation factor eIF-4E. *Neuron* 44: 663-676.

Mercer TR, Dinger ME, Mattick JS (2009) Long non-coding RNAs: insights into functions. *Nat Rev Genet* 10: 155-159.

Mesarovic MD, Sreenath SN, Keene JD (2004) Search for organising principles: understanding in systems biology. *Syst Biol (Stevenage)* 1: 19-27.

Mili S, Steitz JA (2004) Evidence for reassociation of RNA-binding proteins after cell lysis: implications for the interpretation of immunoprecipitation analyses. *RNA* 10: 1692-1694.

Miller MT, Higgin JJ, Hall TM (2008) Basis of altered RNA-binding specificity by PUF proteins revealed by crystal structures of yeast Puf4p. *Nat Struct Mol Biol* 15: 397-402.

Mishra PJ, Humeniuk R, Longo-Sorbello GS, Banerjee D, Bertino JR (2007) A miR-24 microRNA binding-site polymorphism in dihydrofolate reductase gene leads to methotrexate resistance. *Proc Natl Acad Sci U S A* 104: 13513-13518.

Moore FL, Jaruzelska J, Fox MS, Urano J, Firpo MT, Turek PJ, Dorfman DM, Pera RA (2003) Human Pumilio-2 is expressed in embryonic stem cells and germ cells and interacts with DAZ (Deleted in AZoospermia) and DAZ-like proteins. *Proc Natl Acad Sci U S A* 100: 538-543.

Moore MJ (2005) From birth to death: the complex lives of eukaryotic mRNAs. *Science* 309: 1514-1518.

Morris AR, Mukherjee N, Keene JD (2008) Ribonomic analysis of human Pum1 reveals cis-trans conservation across species despite evolution of diverse mRNA target sets. *Mol Cell Biol* 28: 4093-4103.

Morris AR, Mukherjee N, Keene JD (2010) Systematic analysis of posttranscriptional gene expression. *WIREs Syst Biol Med* 2: 162-180.

Mukherjee N, Lager PJ, Friedersdorf MB, Thompson MA, Keene JD (2009) Coordinated posttranscriptional mRNA population dynamics during T-cell activation. *Mol Syst Biol* 5: 288.

Muraro NI, Weston AJ, Gerber AP, Luschnig S, Moffat KG, Baines RA (2008) Pumilio binds para mRNA and requires Nanos and Brat to regulate sodium current in Drosophila motoneurons. *J Neurosci* 28: 2099-2109.

Murata Y, Wharton RP (1995) Binding of pumilio to maternal hunchback mRNA is required for posterior patterning in Drosophila embryos. *Cell* 80: 747-756.

Nicoloso MS, Sun H, Spizzo R, Kim H, Wickramasinghe P, Shimizu M, Wojcik SE, Ferdin J, Kunej T, Xiao L, Manoukian S, Secreto G, Ravagnani F, Wang X, Radice P, Croce CM, Davuluri RV, Calin GA (2010) Single-nucleotide polymorphisms inside microRNA target sites influence tumor susceptibility. *Cancer Res* 70: 2789-2798.

Olivas W, Parker R (2000) The Puf3 protein is a transcript-specific regulator of mRNA degradation in yeast. *EMBO J* 19: 6602-6611.

Ong SE, Mann M (2005) Mass spectrometry-based proteomics turns quantitative. *Nat Chem Biol* 1: 252-262.

Oostra BA, Willemsen R (2009) FMR1: a gene with three faces. *Biochim Biophys Acta* 1790: 467-477.

Orphanides G, Reinberg D (2002) A unified theory of gene expression. *Cell* 108: 439-451.

Paquin N, Menade M, Poirier G, Donato D, Drouet E, Chartrand P (2007) Local activation of yeast ASH1 mRNA translation through phosphorylation of Khd1p by the casein kinase Yck1p. *Mol Cell* 26: 795-809.

Parker R, Song H (2004) The enzymes and control of eukaryotic mRNA turnover. *Nat Struct Mol Biol* 11: 121-127.

Penalva LO, Keene JD (2004) Biotinylated tags for recovery and characterization of ribonucleoprotein complexes. *Biotechniques* 37: 604, 606, 608-610.

Preiss T, Baron-Benhamou J, Ansorge W, Hentze MW (2003) Homodirectional changes in transcriptome composition and mRNA translation induced by rapamycin and heat shock. *Nat Struct Biol* 10: 1039-1047.

Rajasekhar VK, Holland EC (2004) Postgenomic global analysis of translational control induced by oncogenic signaling. *Oncogene* 23: 3248-3264.

Rajkowitsch L, Vilela C, Berthelot K, Ramirez CV, McCarthy JE (2004) Reinitiation and recycling are distinct processes occurring downstream of translation termination in yeast. *J Mol Biol* 335: 71-85.

Ranum LP, Cooper TA (2006) RNA-mediated neuromuscular disorders. *Annu Rev Neurosci* 29: 259-277.

Rinn JL, Kertesz M, Wang JK, Squazzo SL, Xu X, Brugmann SA, Goodnough LH, Helms JA, Farnham PJ, Segal E, Chang HY (2007) Functional demarcation of active and silent chromatin domains in human HOX loci by noncoding RNAs. *Cell* 129: 1311-1323.

Roy PJ, Stuart JM, Lund J, Kim SK (2002) Chromosomal clustering of muscle-expressed genes in Caenorhabditis elegans. *Nature* 418: 975-979.

Saint-Georges Y, Garcia M, Delaveau T, Jourdren L, Le Crom S, Lemoine S, Tanty V, Devaux F, Jacq C (2008) Yeast mitochondrial biogenesis: a role for the PUF RNA-binding protein Puf3p in mRNA localization. *PLoS One* 3: e2293.

San Paolo S, Vanacova S, Schenk L, Scherrer T, Blank D, Keller W, Gerber AP (2009) Distinct roles of non-canonical poly(A) polymerases in RNA metabolism. *PLoS Genet* 5: e1000555.

Schara U, Schoser BG (2006) Myotonic dystrophies type 1 and 2: a summary on current aspects. *Semin Pediatr Neurol* 13: 71-79.

Scherrer T, Femmer, C, Schiess R, Aebersold R, Gerber AP (2011) Defining potentially conserved RNA regulons of homologous zinc-finger RNA-binding proteins. Genome Biol. 12: R3.

Schiess R, Mueller LN, Schmidt A, Mueller M, Wollscheid B, Aebersold R (2009) Analysis of cell surface proteome changes via label-free, quantitative mass spectrometry. *Mol Cell Proteomics* 8: 624-638.

Schmitz-Linneweber C, Williams-Carrier R, Barkan A (2005) RNA immunoprecipitation and microarray analysis show a chloroplast Pentatricopeptide repeat protein to be associated with the 5' region of mRNAs whose translation it activates. *Plant Cell* 17: 2791-2804.

Selbach M, Schwanhausser B, Thierfelder N, Fang Z, Khanin R, Rajewsky N (2008) Widespread changes in protein synthesis induced by microRNAs. *Nature* 455: 58-63.

Shendure J, Ji H (2008) Next-generation DNA sequencing. *Nat Biotechnol* 26: 1135-1145.

Shepard KA, Gerber AP, Jambhekar A, Takizawa PA, Brown PO, Herschlag D, DeRisi JL, Vale RD (2003) Widespread cytoplasmic mRNA transport in yeast: identification of 22 bud-localized transcripts using DNA microarray analysis. *Proc Natl Acad Sci USA* 100: 11429-11434.

Spassov DS, Jurecic R (2003) The PUF family of RNA-binding proteins: does evolutionarily conserved structure equal conserved function? *IUBMB Life* 55: 359-366.

Srisawat C, Engelke DR (2001) Streptavidin aptamers: affinity tags for the study of RNAs and ribonucleoproteins. *RNA* 7: 632-641.

St Johnston D (2005) Moving messages: the intracellular localization of mRNAs. *Nat Rev Mol Cell Biol* 6: 363-375.

Tadauchi T, Matsumoto K, Herskowitz I, Irie K (2001) Post-transcriptional regulation through the HO 3'-UTR by Mpt5, a yeast homolog of Pumilio and FBF. *EMBO J* 20: 552-561.

Takizawa PA, DeRisi JL, Wilhelm JE, Vale RD (2000) Plasma membrane compartmentalization in yeast by messenger RNA transport and a septin diffusion barrier. *Science* 290: 341-344.

Tavernarakis N (2008) Ageing and the regulation of protein synthesis: a balancing act? *Trends Cell Biol* 18: 228-235.

Tazi J, Bakkour N, Stamm S (2009) Alternative splicing and disease. *Biochim Biophys Acta* 1792: 14-26.

Tenenbaum SA, Carson CC, Lager PJ, Keene JD (2000) Identifying mRNA subsets in messenger ribonucleoprotein complexes by using cDNA arrays. *Proc Natl Acad Sci USA* 97: 14085-14090.

Townley-Tilson WH, Pendergrass SA, Marzluff WF, Whitfield ML (2006) Genome-wide analysis of mRNAs bound to the histone stem-loop binding protein. *Rna* 12: 1853-1867.

Ule J, Jensen K, Mele A, Darnell RB (2005) CLIP: a method for identifying protein-RNA interaction sites in living cells. *Methods* 37: 376-386.

Ule J, Jensen KB, Ruggiu M, Mele A, Ule A, Darnell RB (2003) CLIP identifies Nova-regulated RNA networks in the brain. *Science* 302: 1212-1215.

Urano J, Fox MS, Reijo Pera RA (2005) Interaction of the conserved meiotic regulators, BOULE (BOL) and PUMILIO-2 (PUM2). *Mol Reprod Dev* 71: 290-298.

Vasudevan S, Steitz JA (2007) AU-rich-element-mediated upregulation of translation by FXR1 and Argonaute 2. *Cell* 128: 1105-1118.

Ventura A, Jacks T (2009) MicroRNAs and cancer: short RNAs go a long way. *Cell* 136: 586-591.

Vessey JP, Schoderboeck L, Gingl E, Luzi E, Riefler J, Di Leva F, Karra D, Thomas S, Kiebler MA, Macchi P (2010) Mammalian Pumilio 2 regulates dendrite morphogenesis and synaptic function. *Proc Natl Acad Sci U S A* 107: 3222-3227.

Vessey JP, Vaccani A, Xie Y, Dahm R, Karra D, Kiebler MA, Macchi P (2006) Dendritic localization of the translational repressor Pumilio 2 and its contribution to dendritic stress granules. *J Neurosci* 26: 6496-6508.

Vinciguerra P, Stutz F (2004) mRNA export: an assembly line from genes to nuclear pores. *Curr Opin Cell Biol* 16: 285-292.

Wang X, Arai S, Song X, Reichart D, Du K, Pascual G, Tempst P, Rosenfeld MG, Glass CK, Kurokawa R (2008) Induced ncRNAs allosterically modify RNA-binding proteins in cis to inhibit transcription. *Nature* 454: 126-130.

Wang X, Zamore PD, Hall TM (2001) Crystal structure of a Pumilio homology domain. *Mol Cell* 7: 855-865.

Wang Y, Juranek S, Li H, Sheng G, Wardle GS, Tuschl T, Patel DJ (2009a) Nucleation, propagation and cleavage of target RNAs in Ago silencing complexes. *Nature* 461: 754-761.

Wang Y, Liu CL, Storey JD, Tibshirani RJ, Herschlag D, Brown PO (2002) Precision and functional specificity in mRNA decay. *Proc Natl Acad Sci USA* 99: 5860-5865.

Wang Z, Tollervey J, Briese M, Turner D, Ule J (2009b) CLIP: construction of cDNA libraries for high-throughput sequencing from RNAs cross-linked to proteins in vivo. *Methods* 48: 287-293.

Wickens M, Bernstein DS, Kimble J, Parker R (2002) A PUF family portrait: 3'UTR regulation as a way of life. *Trends Genet* 18: 150-157.

Windbichler N, Schroeder R (2006) Isolation of specific RNA-binding proteins using the streptomycin-binding RNA aptamer. *Nat Protoc* 1: 637-640.

Xu EY, Chang R, Salmon NA, Reijo Pera RA (2007) A gene trap mutation of a murine homolog of the Drosophila stem cell factor Pumilio results in smaller testes but does not affect litter size or fertility. *Mol Reprod Dev* 74: 912-921.

Yang E, van Nimwegen E, Zavolan M, Rajewsky N, Schroeder M, Magnasco M, Darnell JE, Jr. (2003) Decay rates of human mRNAs: correlation with functional characteristics and sequence attributes. *Genome Res* 13: 1863-1872.

Yang Z, Edenberg HJ, Davis RL (2005) Isolation of mRNA from specific tissues of Drosophila by mRNA tagging. *Nucleic Acids Res* 33: e148.

Yeo GW, Coufal NG, Liang TY, Peng GE, Fu XD, Gage FH (2009) An RNA code for the FOX2 splicing regulator revealed by mapping RNA-protein interactions in stem cells. *Nat Struct Mol Biol* 16: 130-137.

Yu Z, Li Z, Jolicoeur N, Zhang L, Fortin Y, Wang E, Wu M, Shen SH (2007) Aberrant allele frequencies of the SNPs located in microRNA target sites are potentially associated with human cancers. *Nucleic Acids Res* 35: 4535-4541.

Zhang L, Ding L, Cheung TH, Dong MQ, Chen J, Sewell AK, Liu X, Yates JR, 3rd, Han M (2007) Systematic identification of C. elegans miRISC proteins, miRNAs, and mRNA targets by their interactions with GW182 proteins AIN-1 and AIN-2. *Mol Cell* 28: 598-613.

Zhang L, Hammell M, Kudlow BA, Ambros V, Han M (2009) Systematic analysis of dynamic miRNA-target interactions during C. elegans development. *Development* 136: 3043-3055.

Zhong J, Zhang T, Bloch LM (2006) Dendritic mRNAs encode diversified functionalities in hippocampal pyramidal neurons. *BMC Neurosci* 7: 17.

Acknowledgements

I thank Prof. Michael Detmar for generous support and discussions.

I am grateful to my previous mentors Profs. Daniel Herschlag and Patrick O. Brown at Stanford University for fruitful collaborations and stimulating discussions.

I thank my Ph.D. students Drs. Regula Halbeisen, Alessia Galgano, Tanja Scherrer, Luca Schenk, and Alexander Kanitz, who contributed with their research, ideas and critical discussion to the presented projects of this Habilitation thesis. I also thank all the project and master students that joined my group, and members of the Detmar lab for stimulating discussions.

I was supported by a Long Term Fellowship and a Career Development Award (CDA0048/2005) from the Human Frontier Science Program Organization (HFSPO), and by grants from the Bonizzi-Theler Foundation, the Swiss National Science Foundation (3100A0-112235), and the ETH Zurich (ETH-19 09-1).

This thesis is dedicated to my wife Therese and to my children Mara, Leon and Theo.

Appendix I-VII

The following articles have been published under 'open access' licence terms and can be downloaded free of charge from the journal's website.

Appendix I
Halbeisen et al. (2008) Post-transcriptional gene regulation: from genome-wide studies to principles. *Cell Mol Life Sci* 65: 798-813.
http://www.springerlink.com/content/10614x25092n2120/

Appendix II
Halbeisen & Gerber (2009) Stress-dependent coordination of transcriptome and translatome in yeast. *PLoS Biol* 7: e105.
http://www.plosbiology.org/article/info%3Adoi%2F10.1371%2Fjournal.pbio.1000105

Appendix III
Gerber et al. (2004) Extensive association of functionally and cytotopically related mRNAs with Puf family RNA-Binding proteins in yeast. *PLoS Biol* 2: E79.
http://www.plosbiology.org/article/info%3Adoi%2F10.1371%2Fjournal.pbio.0020079

Appendix IV
Gerber et al. (2006) Genome-wide identification of mRNAs associated with the translational regulator PUMILIO in Drosophila melanogaster. *Proc Natl Acad Sci U S A* 103: 4487-4492.
http://www.pnas.org/content/103/12/4487.long

Appendix V
Galgano et al. (2008) Comparative analysis of mRNA targets for human PUF-family proteins suggests extensive interaction with the miRNA regulatory system. *PLoS ONE* 3: e3164.
http://www.plosone.org/article/info%3Adoi%2F10.1371%2Fjournal.pone.0003164

Appendix VI
Hogan et al. (2008) Diverse RNA-binding proteins interact with functionally related sets of RNAs, suggesting an extensive regulatory system. *PLoS Biol* 6: e255.
http://www.plosbiology.org/article/info%3Adoi%2F10.1371%2Fjournal.pbio.0060255

Appendix VII

San Paolo et al. (2009) Distinct roles of non-canonical poly(A) polymerases in RNA metabolism. *PLoS Genet* 5: e1000555.

http://www.plosgenetics.org/article/info%3Adoi%2F10.1371%2Fjournal.pgen.1000555

Die VDM Verlagsservicegesellschaft sucht für wissenschaftliche Verlage abgeschlossene und herausragende

Dissertationen, Habilitationen, Diplomarbeiten, Master Theses, Magisterarbeiten usw.

für die kostenlose Publikation als Fachbuch.

Sie verfügen über eine Arbeit, die hohen inhaltlichen und formalen Ansprüchen genügt, und haben Interesse an einer honorarvergüteten Publikation?

Dann senden Sie bitte erste Informationen über sich und Ihre Arbeit per Email an *info@vdm-vsg.de*.

Sie erhalten kurzfristig unser Feedback!

VDM Verlagsservicegesellschaft mbH
Dudweiler Landstr. 99
D - 66123 Saarbrücken

Telefon +49 681 3720 174
Fax +49 681 3720 1749

www.vdm-vsg.de

Die VDM Verlagsservicegesellschaft mbH vertritt

Printed by Books on Demand GmbH, Norderstedt / Germany